TRINITY

The Best-Kept Secret

Jacques F. VALLÉE
Paola Leopizzi HARRIS

TRINITY: The Best-Kept Secret

by Jacques F. Vallée & Paola Leopizzi Harris

Copyright ©2021, 2022 by StarworksUSA, LLC and Documatica Research, LLC
All rights reserved.

Published by StarworksUSA, LLC and Documatica Research, LLC

No part of this book may be used or reproduced in any manner whatsoever without written permission except in the case of brief quotations embodied in critical articles and reviews.

ISBN: 979-8-8416-5335-6

First Edition – June 2021 – revised September 2021
Second Edition – August 2022

Cover and interior design by Maryann Sperry
Back cover photo credit: Paola Leopizzi Harris

Published in the United States of America
Printed in the United States of America

This book is warmly dedicated to the primary witnesses for their determination and their sharp observations:

Officer Jose Padilla

Mr. Remigio Baca (in memoriam)

Lt. Colonel William J. Brophy (in memoriam)

Mrs. Sabrina Padilla

Mr. Faustino Myers

and to their families for their willingness to trust us with their testimony.

CONTENTS

FOREWORD: The Owl Bar & Café, October 2017   ix

PART ONE: The Destroyer of Worlds

    One: Jumbo and the Gadget   1
    Two: San Antonito, August 16, 1945   11
    Three: The Padilla Ranch, August 16-20, 1945   25
    Four: The Secrets are Kept   41

PART TWO: Stallion Gate

    Five: First Site Visit, October 2018   59
    Six: Ground Zero   67
    Seven: Second Site Visit, April 2019   75
    Eight: The Secrets are Exposed   85

PART THREE: The Case of the Silumin Bracket

    Nine: The Investigation   103
    Ten: The Device and the Dreams   125
    Eleven: Back in the Lab   135
    Twelve: A Trinity of Secrets   151

PART FOUR: Nineteen Years Later

    Thirteen: Wright-Patterson Air Force Base   177
    Fourteen: The Socorro Synchronicity   185
    Fifteen: Hynek's Flip   193
    Sixteen: Global Patterns, and a Fourth Witness   213
    Seventeen: Another Roadside Attraction   249
    Eighteen: The Fifth Witness   273

CONCLUSION   291

EPILOGUE   309

ACKNOWLEDGMENTS   315

NOTES AND REFERENCES   319

BOOKS AND MONOGRAPHS   337

INDEXES: People Index, Subject Index   343

ABOUT THE AUTHORS   354

FIGURES

1. The scene at the Owl Bar & Café, San Antonio, New Mexico
2. My friend, Ron Brinkley, the last time I saw him, October 2017
3. Map of White Sands with the Trinity Site
4. Assembling "the Gadget" at the Trinity Site, July 1945
5. The shielded tank used by Enrico Fermi to collect explosion samples
6. "Jumbo" containment vessel, on its heavy-duty trailer at Pope, NM
7. Inspecting Jumbo at the Trinity Site in 2019
8. The first witness: Lt. Colonel William J. Brophy, based at Alamogordo
9. Witness Jose Padilla and "Mama Grande" (1856-1940)
10. Paola Harris reviewing data with the Hon. Paul Hellyer, Toronto, 2006
11. Colonel Corso and Paola Harris at a conference in Pescara, Italy, 1998
12. Paola Harris at the crash site with Mr. Padilla in 2016
13. The same location in 2018, with newly-planted poisonous vegetation
14. Detailed map of the Trinity Site, with probable trajectory of the object
15. The enclosure of "Fat Man," the Nagasaki plutonium bomb
16. The Plutonium assembly "master bedroom" at the McDonald Ranch
17. Discussing the case with Mr. Jose Padilla in San Antonito
18. Searching the bed of the arroyo for residual samples
19. On the hillside: reconstructing the object's trajectory
20. The Silumin bracket
21. Mr. Padilla's original drawings of the craft
22. The simple chemical composition of the metal bracket
23. Appearance and terminology in accounts of "entities"
24. Three objects of interest: Analysis of sizes and shapes
25. The Opal Grinder affidavit
26. Ray Stanford's reconstruction of the landing site geometry at Socorro
27. Bill Powers' analysis of the remarkable traces left at Socorro
28. Dr. Hynek's record of the Socorro insignia, and the initial fake
29. Shapes and dimensions in three documented landing traces
30. Sabrina's testimony, February 2021
31. Faustino Myers and Sabrina Padilla as children, circa 1957
32. Witness no. 5: Faustino Myers' testimony, April 2022

FOREWORD

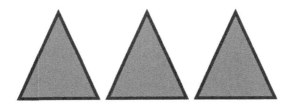

THE OWL BAR & CAFÉ

San Antonio, New Mexico, October 2017

FOREWORD

THE OWL BAR & CAFÉ, OCTOBER 2017

If my friend Ron Brinkley hadn't insisted on buying me a drink at the legendary Owl Bar & Café in San Antonio, New Mexico, I would never have become absorbed in the complex drama of Trinity, and in the forgotten records about the first UFO crash in modern history.

We were driving back after a long, tiring day spent on one of our digging explorations on the Plains of San Augustin (**1**), as we tried to verify some ancient stories of early UFO crashes in New Mexico, and to retrieve evidence of them that we could actually test in the lab. There were many rumors, dating back to the 1940s, when farmers and ranchers claimed to have picked up strange pieces of metal on their land after seeing a weird object in the sky, often right under the noses of Army grunts who boasted of deep secrets and threatened the locals with jail, or worse, if they ever talked about what they'd found.

Secrecy or not, seventy years had passed, and we had in fact picked up a number of interesting pieces of shredded material, undoubtedly from the crash of *something*. I was in a hurry to take them back to a lab in Silicon Valley, where interested colleagues had made plans to run them through the latest testing equipment. I knew the old stories, some of which I had verified, about such items getting lost or stolen, so I felt I was on a mission to secure our data once and for all.

I had a plane to catch, back to San Francisco, and I was a bit nervous. Yet I had known Ron long enough to trust his instinct. It wasn't the first time we had chased elusive witnesses, climbed up and down western hillsides around sites where local people spoke of unidentified lights and scary encounters, and dug up interesting material. Both of us were dusty and tired from the long drive, exposure to the thin air and to the sun at that altitude, so he had little difficulty convincing me that a cup of hot coffee and a piece of apple pie, or a beer with salted almonds, would be a nice break on the way back to the airport in Albuquerque.

I also knew that he must have had something else to tell me; or to teach me.

Over the years I have learned to love the special atmosphere of New Mexico. I had traveled there with Dr. Hynek, back in the 1970s, to work at Corralitos Observatory. I had returned occasionally for meetings of scientists at Los Alamos and other places with a keen, albeit discreet, professional interest in UFOs. But this latest trip had been the most beautiful, an exciting survey of the High Desert, culminating in the recovery of long-buried material samples at a place local researchers had suspected from their own observations, contacts and confidences from trusted neighbors, to be the actual site of a very weird crash.

Ron, the local boy, was proud of his New Mexico ancestry. His family had owned thousands of acres there, raising Longhorns on large ranches, well before White Sands became a huge military base, long before the American West became a modern place with big cities, universities and factories and airports, and lost some of its unique beauty. Yet the mystery remained.

∧ ∧ ∧

"You need to see this place," Ron said as we were sliding into a booth, past dozens of old pictures on the walls, memorabilia from World War Two, bits of letters and many dollar bills stapled to the plaster, fluttering in the air every time someone swung the door open.

"This is where it all started," he went on. "Nineteen-forty-five. Forget everything you've heard about the Roswell crash: It's a very significant story, obviously, but it came two years later; there are many conflicting tales about it, but no tangible evidence remains, and nobody was there to watch it fall. As for the Kenneth Arnold's sighting, which led to the term 'flying saucer,' that was also in June 1947... That's important, but only in terms of what the public was told. Or not told."

"Yes, but the journalists loved it," I said. "People could relate to it."

"The journalists loved it, yes, and the TV people loved it even more," Ron replied, "but in the process everybody had missed the most significant case. It had quietly taken place here, precisely one month after the first nuclear explosion in history, as if in direct response to it. Nobody spoke about it for many years. It happened just a few miles from this diner. That's where we've got to start."

The waitress walked over and took our order. The place was quiet. Local customers were drinking at the bar, relaxing after a day's work. Ron rested his back against the wooden booth and gestured to the open space between us and the other tables.

"Back in 1945, the scientists of Project Manhattan lived around here, in San Antonio, at the critical time for the assembly of the Bomb," he said as quietly as if he'd spoken about his neighbors in some sleepy little village. "If you were an American scientist yanked out of your university campus by the Army, or a Nobel prize-winning atomic physicist from Europe working hard to defeat Hitler, this was the only half-decent place for a meal."

I tried in vain to imagine the scene. Some of the photographs were framed; they showed the big propeller airplanes of the time, men in fatigues standing in front of a hangar. The pictures were old and had turned brown. There were cartoons too, and posters encouraging discretion about the war, stern warnings against rumors and gossip. I took it all in. Or rather, it took me. I knew the historical facts. But it seemed useless for me, a Frenchman born in 1939, to reconcile those few pieces of paper, the faded pictures, the crude drawings, with the cataclysms of the Atomic Age, and this funny watering-hole in the tiny town of San Antonio, in the middle of an America desert.

∧ ∧ ∧

I remember the Second World War. How could I not? The French town where I was born was a key lock on the great Oise River and the only direct road from Normandy to Paris. It was seized by the Germans in 1940, and three years later it became a prime target for the British and the Americans in the days leading to the Normandy invasion, because they needed to cut off German reinforcements. Pontoise, as the town was called, was bombed eighteen times in raids of the Royal Air Force and the US Army. Much of it was reduced to smoking ashes.

Our town was liberated on August 30, 1944.

To a five-year-old child, a world war is surprisingly uncomplicated: The first house I remember, rented by my parents, sat on a hill overlooking the river and the two strategic bridges. It was blown up by a direct hit from a stray American bomb in the middle of an inferno. Our second rented house, further away, also sat on a hill from which we could see the valley and even Paris in the distance, on a clear day. By then I was old enough to watch the aerial battles with my parents when we knew there was no viable shelter: Relentlessly, German batteries were firing at the attackers. Planes fell apart, spiraled down in flames. Snipers took aim at the parachuting pilots, helpless in the sky. That was faster and cheaper than taking prisoners. There was nothing there that a five-year-old kid couldn't watch, hear and understand readily; and remember, too, in vivid detail, although I was very fortunate not to see death at close range.

It was combat, but it was "normal" combat, in a classical war, fought with bullets and chemical explosives. I saw what that could do. But neither I nor my parents had any concept of a nuclear war. Even today, there's never been a nuclear explosion in Western Europe.

Now I was thinking of far-away France while eating some apple pie in a quiet diner in New Mexico and my friend Ron was reminding me that we were just a few miles from the Trinity site, the location of the first atomic blast in history.

My mind sank into the scene; it drifted to the horror and the splendid immensity of it.

Ron misunderstood my silence. As if upset that I wasn't paying enough attention, he leaned over the table and insisted: "That old house, ranch-style, dilapidated, the one we passed, across the street, that's where Dr. Robert Oppenheimer stayed at the time (2). He'd come in here, always with a briefcase full of papers. I've spoken to the old-timers: They greeted him and just called him 'Oppie.' They had no idea what he really did. But they knew enough to leave him alone, and not to ask questions. He would sit in that corner, away from the GIs and the cowboys, sometimes with colleagues who came from Oak Ridge and Chicago, to work on the equations for the bomb. Some had escaped from Italy, Poland, Hungary, Czechoslovakia, Germany, all over."

Ron concluded, calmly: "This place is legendary. It changed the world."

I trusted Ron, his sense of history, his passion for details about the lives of local people he knew so well. Another reality was slowly painting itself in my mind as I listened to his recounting of the four weeks that changed everything.

He took a drink, leaned forward, all business now:

"The first test took place on July 16, 1945. It was a plutonium bomb, hanging from a tower at the place that became known, for future history books, as *Ground Zero*. That's when the military knew they were ready. Three weeks later, on August 5th, (August 6th in Japan), "Little Boy," a novel but relatively simple uranium bomb, reduced Hiroshima and its inhabitants to fine radioactive dust. And just three days later, on August 8th, a plutonium bomb they called "Fat Man" in homage to Winston Churchill pulverized the giant military naval complex and the city of Nagasaki. On August 14, stunned and disheartened, the emperor of Japan, the living Son of Heaven, capitulated without conditions. *And two days later, August 16th, the first UFO crash in modern history took place a few miles from here."*

"The funny thing is, nobody has done any serious research about it. With a few exceptions, the folks who call themselves 'ufologists' have remained fixated on Roswell."

"Why is that?" I asked.

"Because the Army covered up everything very well and the two witnesses of this particular crash were too scared to speak up. The scientific community was kept in the dark, as usual. There were good reasons for that."

So, I finally realized, that's why Ron had insisted on grabbing a drink at the Owl Bar & Café. We had worked together long enough for him to know I would take the hint, and explore the trail he'd mapped out for me.

∧ ∧ ∧

As a researcher of ufology, I have been criticized for not taking an active enough interest in contact and "crashed saucer" cases, like Roswell.

The accusation is a valid one.

As a young scientist in Europe, I had read the books of "Professor George Adamski of Mt. Palomar observatory" in French translation with amused interest, relieved by occasional giggles when he spoke of his scientific prowess (he had a small telescope on the slopes of the mountain) and his extrasolar paramours **(3)**. From the admittedly inflated attitude of educated Europeans, it was easy to dismiss such American saucer tales, especially because American scientists were doing the same thing. Another early "contactee" advocate, Frank Scully **(4),** was more credible but his technical backing quickly evaporated, and again I was left with amused skepticism.

All of us, as it turns out, were wrong about Scully, but that's another story.

On June 21, 1947 there was an incident at Maury Island in Puget Sound, in the State of Washington, considered the first American crash case complete with "evidence" and multiple witnesses. It came just before the celebrated sighting by pilot businessman Kenneth Arnold and seemed to support its credibility, yet it only yielded a pile of slag and a lot of welcome publicity for entertaining pulp, but no worthwhile data. My colleagues assure me it was a hoax. I am not so sure. I regret not following up on that one. Today I would still like to get a little piece of that slag, and test it, just in case.

Fig. 1: The scene at the Owl Bar & Café, San Antonio, New Mexico. (Photo: JV)

Foreword

Fig. 2: My friend, researcher Ron Brinkley, the last time I saw him, October 2017.

When I moved to the US with my wife in 1962, we brought with us the first consistent international computer-savvy database of vetted UFO cases to Austin, Texas, where I worked on the first digital map of Mars for NASA and recorded galactic observations at McDonald Observatory while pursuing the statistical study of the UFO phenomenon: After eliminating such known sources of errors as the planet Venus and the occasional swamp gas, definite patterns emerged, in space and time. This led us to correspond with Dr. Hynek, the Air Force's scientific consultant on Project Blue Book (**5**). In 1964 I joined him at Northwestern University to work on a doctorate and pursue this research. Crashed saucers were not on the agenda, but Allen had read all those reports. He showed me why there was something wrong with every story. Either witnesses were missing, testimony was dubious or second hand, or vital evidence simply could not be found, no matter how hard you looked: there was nothing for a scientist to take to the lab, and secrecy wasn't the only reason, or even the main one.

Some stories about a "special weather balloon" or a cluster of high-altitude *radiosondes*, or test dummies with parachutes, were offered as "explanations." Everybody got confused.

Back at Wright-Patterson Base in Ohio, which I visited with him in 1964 on a special, two-day clearance, we were shown a metal cabinet with many carefully-labeled pieces of strange rocks and ordinary twisted metal that the American public had sent to the Air force as... that long-tortured, misused, misplaced word again: *Evidence!*

Even after researcher Stanton Friedman and others seriously investigated and publicized the Roswell case of early July 1947, with its welcome flurry of testimony, field research yielded nothing more than proof of blatant, continued, boring official lies.

Evidence of official lying about UFOs is not evidence of the reality of UFOs.

Deceptive official posturing at the service of bureaucratic tranquility is nothing new, even in the form of lies to Congress, under oath, in uniform, your right hand on the Holy Bible. The few scientists who doggedly followed the reports, interviewed witnesses and compiled serious files could only look at each other, shake their heads, and ask: *"If you were smart enough to come here from Alpha Centauri, why would you crash on that guy's farm?"*

I still don't have a good answer to that one. But what if you didn't come from Alpha Centauri at all? What if your spacecraft *was designed to crash*? What if it was a gift? Or a signal? Or a warning? The hopeful inception of a strategic conversation? What if it wasn't a "spacecraft" in our current, primitive sense of the word? What if you didn't care if the occupants died? Nobody had seriously considered those alternatives.

∧ ∧ ∧

Once I was established in the US, with a PhD in artificial intelligence and a job in real science, I did make renewed efforts to probe some of the stories. Maintaining a critical, even skeptical bias, I began a collection of artifacts that has expanded to some twenty-five items, mostly (but not always) of metallic composition, from four continents. But such a collection, like the one at "Wright-Pat", doesn't prove anything by itself. So I sought out the early believers and even the cultists. I met with George Hunt Williamson, one of Adamski's supporters, who left me puzzled. He had signed the affidavit stating he was a witness to a "Contact" Adamski claimed in November 1952, but he admitted to me he'd not *actually* seen an Alien, or even a saucer. I went to the site, in the Mojave Desert, to

Foreword

understand the lay of the land: Maybe he had glimpsed a silhouette, and some sort of light in the sky. (**6**) I came back empty-handed.

As my research in computer science took me to Stanford University and to SRI, I contributed to the development of social networking on the early Arpanet and the later Internet as one scientist among our group of Principal investigators for the Defense Advanced Research Projects Agency (DARPA). That work brought me into contact with government teams that had their own files—and considerable yet discreet interest—regarding the UFO problem at the highest level. Those contacts expanded when my colleagues and I developed a series of venture capital funds to invest in medicine and high-tech, an activity that continues today. Alongside this professional activity, I continued to research the history and structure of the UFO phenomenon.

During most of that period, government data about recovered craft and purported Alien bodies was highly classified, so I tried to find corroborative evidence elsewhere. In Chicago, I'd spoken to Ray Palmer, a remarkable man, the genial editor of *Amazing Stories* and of the early pulp magazine *Flying Saucers,* who had launched the first reports and financed Kenneth Arnold's later investigations. He could only repeat stories I'd read a dozen times. They all began in the summer of 1947.

If there was a genuine UFO crash in 1945, two years before Roswell, but only four weeks after the first atomic explosion, that was an extraordinary fact.

For one thing, it meant that practically all the books about UFOs, whose first sentence is always: "The Flying Saucer Era began on June 24, 1947 when businessman Kenneth Arnold saw an unidentified object in the sky..." were plain wrong. Arnold's sighting is reliable and it triggered the media's fascination with the problem, but the 1945 case was more important. It represented a source of key questions one could no longer avoid. The first one being, "Why has everyone missed it?" The second one: "Are some witnesses still around?" The third one: "Why on Earth would they remain silent all those years?" And the fourth one: "Where's the damn Evidence?"

Those are questions that knowledgeable Italian journalist Paola Leopizzi Harris had started to ask before me, and even before Ron. Formally trained in Italy as an educator, and a long-time, trusted research associate and translator for Dr. Hynek, she had been at the site before any other investigator. She is the only researcher who met both witnesses, and recorded their testimony. She wrote to me at the end of July 2018 when she became aware of my interest in the case, and we made arrangements to meet in September when she came to lecture in California.

Paola told me about her frustration to receive no support to analyze the data: "I have worked on it for seven years and the only people to take it seriously are outside the country," she said. "American researchers do almost no field work, arguing it's expensive and takes too much time! I can bring you a slice of metal from a bracket, retrieved from inside the craft by the nine-year-old boy who took it from a plaque on the wall."

When I realized how much work she had done, and we spoke about how much more could be accomplished, we joined forces and quickly assembled a small team to re-examine the facts and further research the above four questions. You will see how they turned out to be intertwined, in spite of their extraordinary and occasionally shocking nature, in remarkably rational ways.

We were able to take her initial observations to the next logical level. But that investigation itself opens up even deeper questions.

Our findings are the subject of this book. They challenge the very nature of this field of research and yes, of science. They pose fundamental interrogations not just for a few scientists, but for Humanity itself: its past history; its present and its future.

In the weeks that followed that congenial discussion at the Owl Café, I began assembling the documents that had randomly accumulated in my own files, about related events of the time: folders with names like Alamogordo, Socorro, Los Alamos, San Antonio (the town in New Mexico, not the one in Texas), White Sands, and all the notes I had taken over the years, after meetings with colleagues from NASA Ames, Lawrence Livermore, Oak Ridge and Brookhaven, atomic laboratories I had visited in the course of my scientific work and my computer career; all major research centers where there were active government scientists willing to talk to me, always off the record, about what they had seen, and done.

Other researchers who had explored the Southwestern sites before me graciously contributed their own research documents. Paola compiled vast repertories of notes, films and tape recordings for us to transcribe and review. They painted a stunning background for a true history of the UFO saga.

Foreword

The picture that emerged was of a dangerous time, a period of seemingly super-human creativity, of science at the extreme edge of secrecy and danger, of crazy experimentation with elements unheard of in all of history, without any sure guidelines for controlling their release, and indeed, no moral precedent to assess the scale of the destruction that would follow, or the waste in thousands of human lives.

Ron Brinkley's blunt history lesson at the Owl Café, one day in October 2017, led to the research that resulted in this book. The events we are about to describe, along with the step-by-step details of our investigation with its twists and turns, demonstrate the existence of levels of reality science has failed to recognize, even as it unlocked the power of the atom: *somebody was watching*. From that side of reality came material forms we could perceive, decode, and partially understand, yet the phenomena were beyond human technology at the time, and they remain an enigma today.

Who are these emissaries from elsewhere? Once we set aside the deceptive mythologies that have accumulated around the problem—silly tales of Martian bases and political delusions about Nazi bases at the South Pole, speculation about superior races and the obsession of complicated conspiracies by government insiders—we are left with plain evidence for an unknown intelligence.

Why should it come as a surprise that consciousness can take larger forms than our limited brains have imagined? And how can we start designing an appropriate research program capable of interacting with other manifestations of life in a fantastic universe we've only started to map?

I dearly wish I could share the thrill of the resulting quest and our early conclusions with Ron, but we lost him in August 2018. He was on his bike, at five in the morning, rushing to his job at an airport boutique that sold local Indian art to tourists in Albuquerque. A speeding truck hit him hard, dragged his body and crushed his head, in spite of the helmet he was wearing.

Ronny, we miss you. The research you inspired will continue. But I will never drive across the range again without thinking of what you taught me about the beauty of that land, its people, their history, and the mysteries they remember.

PART ONE

THE DESTROYER OF WORLDS

*Now I am become Death,
The Destroyer of Worlds.*

Bhagavad Gita, cited by Dr. Robert Oppenheimer, 1945

CHAPTER ONE

JUMBO AND THE GADGET

"Laura, don't be afraid of becoming a widow. If Enrico blows up, you'll blow up too..."

What a thing to say to the wife of a friend!

Yet this was the solemn, starkly realistic statement of Italian physicist Emilio Segré to Laura Fermi when he visited atomic scientist Enrico Fermi and his family in Chicago in 1943. Laura understood perfectly what he meant.

At the time, Nobel laureate Fermi **(7)**, who had fled fascist Italy in 1938, spent most of his time under the bleachers of the stadium at Stagg Field, three blocks away from their home, conducting the secret work on the first atomic reaction "pile." The first controlled nuclear chain reaction in history had taken place there, on December 2, 1942 under his direction.

In complete secrecy.

Two years later, the science work had moved to New Mexico, and the Fermi family lived in Los Alamos, where Laura worked as a medical assistant to Dr. Louis Hempelmann, of the Los Alamos Health Group.

"Early in July (1945)," writes Laura in her intensely-personal memoir *Atoms in the Family* **(8)**, "men had started to disappear from the mesa and the word *Trinity* had floated with insistence in the air."

By July 15, 1945, only women remained in the town of Los Alamos. That afternoon, friends of Laura told her they were about to drive to the Sandia Mountains, near Albuquerque, some 140 miles due north of the secret testing site they knew as *Trinity*, and spend the night there, in hopes of "seeing something."

The site had been selected in part because it was within the Alamogordo Bombing and Gunnery Range, established in 1942. The appropriately called *Jornada del Muerto* ("Journey of the Dead") was chosen for its isolation in terms of both secrecy and safety. Yet, as an Army brochure wrily notes, "it was still close to Los Alamos for easy commuting back and forth."

One has to remember that a round trip of nearly 300 miles in a scorching desert is considered as a short escapade by the standards of the American Southwest. The Army had surveyed seven other sites in California, Texas, New Mexico and Colorado before they picked Trinity.

Soldiers had begun assembling at the site in the fall of 1944, and the population grew throughout 1945 with the arrival of technical personnel. They would be greeted by a flat plain of poor soil where few trees grew among thousands of acres of mesquite and creosote-covered sand, with the occasional rattlesnake as rare entertainment.

On the southeastern horizon, vaguely delineated, was the grey-blue mass of the Malpais Mountains, forming a background before which stood three smaller peaks: *Oscura* (which rose to 8732 feet), *Little Burro* and *Mockingbird*, on a 20-mile diagonal line segment from northeast to southwest.

Those three peaks had given the area its beautiful name of Trinity, soon to come into historical records as the symbol of the world's introduction to the Atomic Age. Trinity also marked the end of innocence, if there ever was such a thing, for America--and the rest of Humanity.

∧ ∧ ∧

On the morning of July 16, 1945, Laura Fermi heard talk of a "blazing" light seen at dawn by one of the male patients at Los Alamos hospital, who was awake at 5:30 in the morning. She was eager to learn what had taken place, and why such an unearthly light had preceded the Sun.

A few of the men came home in the evening, exhausted, their skin dry and seemingly hardened. Enrico Fermi was among them, so tired he staggered home and collapsed in bed. All he said the next day was that

Chapter One - Jumbo and the Gadget

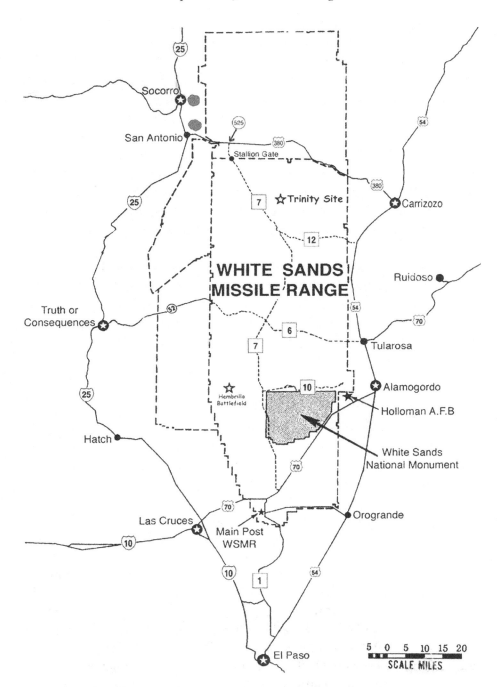

Fig. 3: Map of White Sands, the Trinity Site and two UFO close encounters
Note that Los Alamos is located 150 miles North of Socorro.

he'd been unable to drive, and got a friend to take the wheel. A local paper mentioned the extraordinary light, so bright even a blind girl had noticed it.

Perhaps an ammunitions depot had blown up? People were disciplined enough to go back to work without any more gossip. But a neighbor, Genia Pieiris, who always managed to learn about things before anyone else, took Laura aside and whispered, "Listen, at Trinity, they must have exploded... an atomic bomb."

Enrico Fermi, that morning, had been among the first scientists at the actual site. He had done a preliminary evaluation of the energy from the blast. His rough computations over, he climbed into a Sherman tank covered with radiation-absorbing lead and collected samples in the half-mile depression the 19-kiloton plutonium bomb had dug up in the desert (Fig. 5).

Brigadier General Thomas Farrell has described the sight:

"The effects could well be called unprecedented, magnificent, beautiful, stupendous and terrifying. No man-made phenomenon of such tremendous power had ever occurred. The lighting effects beggared description. The whole country was lighted by *a searing light with the intensity many times that of the midday sun.*" **(9)**

^ ^ ^

History rolled on from there, implacable and deadly. Following that successful "test" of a bomb simply called "The Gadget," containing six kilograms of plutonium, things happened very fast. The Pentagon was tired of fighting a horrible war of attrition against what was left of the great Japanese military might, in desperate battles with high losses on both sides, on every single God-forsaken little island across the Pacific. Also, they must have suspected that German physicists were getting close to testing a similar war-ending, history-making bomb. **(10)**

On August 5, 1945, a simple-design uranium device nicknamed "Little Boy" was dropped on Hiroshima by the crew of the Enola Gay. President Truman lost no time in revealing its nature to the world. And three days later, on August 9, 1945, as Ron had reminded me, the 10,800 pound plutonium-based "Fat Man," based on the design of "the Gadget" tested at the Trinity site pulverized the sophisticated military complex at Nagasaki harbor, the pride of Nippon heavy industry, the heart of its invincible Navy. In the process, it wiped out the munitions plant that had manufactured the torpedoes used to sink the American ships at Pearl Harbor.

Chapter One - Jumbo and the Gadget

Fig. 4: Assembling "the Gadget" at the Trinity Site, July 1945.

On August 15, 1945, the Emperor of Japan, broken, humiliated and confused, issued an *Imperial Rescript of Surrender* accepting the terms of the Potsdam Declaration, and on September 2, 1945, his envoys signed the *Instrument of Surrender* on the flight deck of the USS Missouri.

And two days after the Emperor's surrender, just one month after the awesome atomic test by Robert Oppenheimer, Enrico Fermi and their star team of physicists, a weird object came out of nowhere, circled in the sky over the mesquite-infested hills only some 10 miles North of Trinity, and crashed on a New Mexico ranch, as if to underline the enormity of what had taken place; or to acknowledge it?

Or to offer a warning perhaps, a sign that somebody else had been watching?

Or even, in my own speculation, to open a parallel historical and strategic path, one that superseded even the taming of the atomic force?

Fig. 5: The special shielded tank used by Enrico Fermi to collect explosion samples.

Fig. 6: "Jumbo" containment vessel, on its heavy-duty trailer at Pope, New Mexico.

Chapter One - Jumbo and the Gadget

Is that why, a few years later, Canadian government scientist Wilbert Smith, of Project Magnet, would be quietly briefed on US research on UFOs and told it was classified "higher than the atomic bomb?"

^ ^ ^

Let's pause here for a while.

It is important, before we move on, to really understand what took place at Trinity, how it was accomplished, and why it ushered in a new era for the human race: suddenly closing one door to conventional war and centuries of tribal and national politics, and opening wide another door to a new world that was no less dangerous, complex and ultimately terrifying.

The Manhattan Project, formed in June of 1942, had been given secret responsibility to develop an atomic bomb. It was already clear, from Intelligence information and from confidential reports from networks of European scientists gathering information on German projects (such as the group created by my late publisher and friend Jacques Bergier, one of the earliest French atomic spies, who was involved in the tracking of "Heavy Water" in Norway and in the discovery of the Peenemunde rocket base) that the Nazis were seriously working towards a nuclear breakthrough.

The Manhattan Project implemented three enormous secret facilities for specific science goals: Gas diffusion and electromagnetic processes at Oak Ridge, Tennessee; nuclear reactors capable of producing the necessary plutonium at Hanford, Washington; and the design and actual construction of the bomb at Los Alamos in New Mexico, under Dr. Robert Oppenheimer.

In order to calibrate the radioactivity measuring instruments, a preliminary test was done on May 7, 1945 at the Trinity site. It was not a nuclear weapon, but rather a dirty bomb that used a conventional explosive with a 100-ton power that was mixed with materials obtained from Hanford, and dispersed the radiation over what the Army casually called "a fairly wide area."

In the meantime, Dr. Oppenheimer's team developed two designs for a real atomic bomb: a simple one using Uranium 235, which scientists were confident wouldn't require preliminary testing; but also the more complex plutonium scheme, where lens-shaped charges of conventional explosives would all explode at the exact same millisecond to compress a plutonium ball into a critical mass to trigger a nuclear chain reaction. A scientific first.

Fig. 7: Inspecting "Jumbo" at the Trinity Site in 2019. (Photo: Paola Harris)

The plutonium effort was complex because the design had to encompass not one explosion, but two: First, the conventional blowing up of the TNT charges to compress the core; next, the actual chain reaction in which the atoms themselves split, releasing enormous energy and neutrons that, in turn, strike and split more atoms...

... all in one millionth of a second.

There was one major risk: if the chain reaction failed to start, the TNT would shower the New Mexico landscape with highly radioactive plutonium dust. Not only would this waste some extremely rare plutonium but it would endanger all life within a significant part of the United States for centuries.

So plans were made to test the explosion inside a "safe" enclosure, a vessel so large, so thick and so strong that it could contain the first explosion and allow recovery of the plutonium if the chain reaction failed to materialize.

They called the big testing device "Jumbo."

They designed an extraordinary steel chamber that was 25 feet long, 12 feet in diameter and weighed 214 tons. Manufactured in Ohio by the firm of Babcock and Wilcox, Jumbo was brought to Pope, New Mexico by a dedicated train.

It was moved to the Trinity site on a specially-built trailer with 64 wheels (Figure 6). The trailer was hauled over by two Caterpillar DC-7 tractors, one pushing and one pulling. The Army built a special raft to take "Jumbo" across the Rio Grande.

A modern Army brochure (**9**) mentions that in the end, Jumbo was never used in the test because the scientists gradually became confident that the plutonium bomb design would work: Why waste some precious material? Then on July 12, 1945, the two hemispheres of plutonium from Hanford were brought to the McDonald Ranch for assembly.

Asked to sign a receipt for it, Brigadier General Tomas Farrell recalls demanding to hold and handle it:

"I took this heavy ball in my hand and I felt it growing warm. I got a certain sense of its hidden power. It wasn't a cold piece of metal, but it was really a piece of metal that seemed to be working inside. Then, maybe for the first time, I began to believe some of the fantastic tales the scientists had told me about this nuclear power."

Left under a steel tower only 800 feet from Ground Zero, Jumbo survived the exposure to the first nuclear blast when "Gadget" was detonated (the steel tower didn't). In the end, Jumbo was only used for tests of non-nuclear explosives the following year. One sizeable steel section now rests, preserved on a pedestal downtown Socorro, the accident dignified with an explicatory plaque, after a somewhat careless, tumbling flight.

Jumbo was never used for a radioactive test. It was only called into service once, a year later, when the Army detonated eight 500-pound conventional bombs inside it. Because the device was standing vertically, an official brochure states, the bombs were stacked at the bottom, resulting in both ends of the cylinder being blown off, as I could see during a site visit (Fig. 7). Today, what remains of Jumbo still lies in the desert sun. It's not going anywhere. With its 25 feet length and 12-foot diameter, it is a permanent witness to the enormity of the human and scientific challenge the military had undertaken at that forsaken place.

A short distance northwest of the Trinity site, hundreds of ordinary citizens of San Antonio, New Mexico were about to experience the same "hidden power" that so impressed General Farrell. But they were never warned about it.

CHAPTER TWO

SAN ANTONITO, AUGUST 16, 1945

Barely 18 miles northwest of Trinity stood the ranch home of the Padilla family, described by researcher Ryan Wood in his 2005 book *Majic Eyes Only* as "a traditional adobe house with a well in the front yard and a nice shiny pickup truck that (nine-year-old son) Jose was allowed to drive, since he was the only (young) male in the family not serving in the armed forces at the time." (**11**).

Jose's dad, Faustino Padilla, worked for the Federal Refuge in a development, *El Bosque Del Apache* near San Antonio, at the W.P.A (Work Projects Administration of the Civilian Conservation Corps).The family supplemented their income by raising cattle on a large ranch leased from the Bureau of Land Management, and much of the everyday work on the property was delegated to Jose, often aided by his friend "Reme" (Remigio) Baca, who was two years younger.

Living so close to Ground Zero, the families of Jose and Reme would soon experience a sudden change in their world, and the two boys would be the first witnesses to the extraordinary event that marks the true beginning of the UFO era in the United States—still unexplained today in spite of its obvious repercussions in the current debates. To the Army, the event would be just one more local mission to investigate the destruction of a radio tower and to recover a strange craft fallen from the sky during a thunderstorm. The young soldiers in the unit barely realized where they were, the history of the place, or the long traditions of the people who actually lived there.

Nor did they care: they came in with a Jeep and a big truck, did what they were told to do and went back to White Sands a few days later. To us, however, analyzing the events in the light of 75 years of accumulated data, what the two boys experienced at that special location needs to be brought to light and analyzed in the full context of American technology and culture at the critical time of the end of the Second World War.

∧ ∧ ∧

The family of Reme Baca lived in conditions very similar to those of Jose Padilla: Remigio Baca's father, who had first worked as a sharecropper, later became an attendant at Veterans Hospital in Albuquerque. Both men were also occasionally employed by none other than Conrad Hilton, who owned several businesses in San Antonio, including the very first Hilton Hotel, and by a Mr. Alliare, who managed a mercantile business there.

The Baca family members were long-time settlers of the area, descendants of the original Spaniards who established the government of New Spain. Remigio ("Reme") traces his family roots to Captain Cristobal Baca, a native of Mexico City who settled in New Mexico with his wife in 1600. Their descendants would become prominent residents of the State, while Jose Padilla traces his ancestry in part to more recent Spanish immigrants, but also to the long history of his Native American lineage through his mother's family.

The initial inhabitants of the land were composed of three Native groups: "the Mogollon, who lived in the high lands of New Mexico and Arizona; the Hohokam, farmers of the southern deserts, and the pueblo-dwelling Anasazi of the high mesas to the north," noted Reme in a memoir entitled *Born on the Edge of Ground Zero* (**12**). But these tribes were already in decline when the Spaniards arrived and built their own settlements, while five tribes of the Apache People spread their power throughout the area, and into Northern Mexico, often with bloody struggles.

Danny Many Horses, a modern artist, friend of Paola and keeper of the Mescalero Apache tradition in the Southwest, gave me precious background on the troubled history of the region when it attracted first the Spaniards from Mexico and the white settlers from the East. Finally, in 1743 a Spanish leader agreed to designate areas of Texas for the Apaches to occupy, easing the battles over land. A Wikipedia entry recalls that "In a ceremony in 1749, an Apache chief buried a hatchet to symbolize that the fighting was over, thus the term we use today, 'bury the hatchet.' As time went on, the Apache Indians developed a strong bond with the white men of the area. At first, relationships seemed solid, and the Apache felt protected.

As things progressed, however, raids by the settlers started again; they included the slaughter of Native American people and the theft of their goods and livestock.

At the end of the Civil War, after 1865, violence became even worse. Many defeated Confederate soldiers came west to seek new fortunes. Some went on to California and Nevada where gold had been discovered in 1849, while others joined the US Army's battles against the Indian population. It was a horrible time, since one of the Government's favorite tactics was to kill all the buffaloes in an area to deprive the tribes of their traditional source of meat.

The Apaches were nomadic warriors; they formed alliances with the sedentary peoples and protected them, but the hired white marksmen kept killing animals all over New Mexico and Arizona, decimating the herds, starving the local inhabitants. Many Apaches retreated to their mountain hideouts and sought the wisdom of their Elders. They saw multi-colored orbs there, "not just lightning," Danny Many Horses told me: "The lights meant that Ussen, the Giver-of-Life, the Omnipotent God, was watching over them. He told them that hideous creatures would come down from the clouds if the tribes didn't live in harmony. There were spirits in the moon, in the stars: The Apaches could hear their grumbling, high above the mountains."

In time the local people, including the Hispanics and members of the Pueblo and other tribes, were able to live in peace with the white settlers again. Among them was the Chiricahua Apache Tribe, from which Jose Padilla is descended on his mother's side: His great-grandmother was none other than "Mama Grande," Maria Amada Chavez, (1856-1940) who was a local Chiricahua Apache leader. Mama Grande was recognized as the community authority in the area, where she is remembered as a wise and powerful woman. The community of San Antonito was named after her son, "Little Antonio."

As of 1940, there was a record of only 35 Apache Indians living in the state of Oklahoma, but in 1970 a record of about 1,500 were documented in New Mexico.

On the side of Jose's father, in contrast, the Padilla family is of recent Spanish origin: His grandfather Trini came from Barcelona and his father Faustino was born in February 1890 in Franklin, Texas, a community that became part of El Paso.

As for the 1945 crash site, with its convenient geography and access to water, it was a favorite place for the tribes when they moved across the region, pitching their tents and watering their horses. Evidence has been found of the burnt, hollowed-out logs they used as water pipes, and cultural archaeologist Kay Sutherland has theorized that the "painted images" of the Mogollon, preserved on stones surfaces, may have been used "as a mystical gateway to the spirit world."

Mystical or not, the place was destined to serve as an interface with unknown realities.

∧ ∧ ∧

"Inez Padilla, Jose's mother, had just seen her husband Faustino off to work," recalls Reme Baca, "and Jose Padilla was drinking a cup of hot chocolate when the flash of the plutonium bomb, the sustained deadly sound and the heat wave came rolling over the landscape. Surprised, his mother peeked through a crack in the door at the flash of light. As a result, Inez sustained permanent loss of sight in that eye." (**12**)

Jose sought cover when the explosion came, and he recalls the hot wind of the blast engulfing the house three minutes later.

Reme Baca himself was seven at the time. He mentions being awakened by his bed shaking, as if jumping, as a bright light came into his room. "It seemed like a train was coming through the front door," he recalled. The shock wave broke windows as far as 120 miles away.

In the following days and weeks people would realize that what they had seen and heard wasn't simply the result of an explosive test, or an accident in a munitions storage area at the Alamogordo Bombing range, as the Army had initially announced. But the full story wouldn't come out until after Hiroshima on August 6th, when President Truman made the formal announcement to the world.

With the Nagasaki strike on August 9th and the negotiations for the surrender of Japan, everybody finally started breathing more freely, knowing that the conflict was coming to an end, and soldiers would soon return to their families and to their jobs.

Chapter Two - San Antonito, August 16, 1945

Activity around San Antonio gradually seemed ready to return to normal.

Nobody saw the object arrive from the sky.

∧ ∧ ∧

Although the case has largely been neglected, or cursorily noted and then forgotten in all the sensational television arguments about Roswell, a few investigators did take an interest in the story. Timothy Good, a careful chronicler of modern ufology, tersely mentioned the crash in his 2007 book *Need to Know*:

> In August 1945, (…) two young boys, Jose Padilla and Reme Baca, were doing various chores on the Padilla Ranch at San Antonito, near San Antonio, New Mexico, when they saw a brilliant light, accompanied by a crunching sound. Eventually, they came upon a long gouge in the earth, at the end of which lay a dull grey avocado-shaped craft, with a protuberance towards the end, which could be seen through the smoke from its burning wreckage. Scattered among the debris were pieces of shiny tin-foil-like metal. A piece that had become lodged under a rock unfolded itself instantly. It was kept at the ranch home for a number of years. **(13)**

Timothy Good gives as his source "personal interviews in Seattle" (near where Reme Baca had gone to live) performed on June 6th and 7th, 2004, and "numerous subsequent communications with both witnesses." He obviously did his homework, but he then went on with his other investigations.

There was one notable exception to the benign neglect that surrounded the case: In long, detailed interviews that Italian investigative journalist Paola Harris did on July 5, 2010 after she'd tracked down Reme Baca to his home in Gig Harbor, Washington State, she understood why the case filled important gaps in the story of recovered unidentified craft. **(14)**

She completed that interview with a detailed phone conference with Jose Padilla, then a State Trooper in Rowland Heights, California.

Working independently, guided by her extensive knowledge of the phenomenon around the world, and perhaps by a sense of history that is felt more acutely by Europeans, she grasped the deeper importance of the case as the very first UFO crash in the recent American record, fully two years before Roswell. It wasn't well-known, simply because the initial story hadn't been followed up by publicity-seekers; and perhaps because other, more obscure reasons made it particularly challenging. So she went to work, sponsored technical teams to help her, quietly tracked down related information and started digging into the details.

Television reporters followed her to the site with their cameras. Among them was the dedicated California videographer James C. Fox (director for the documentaries *I Know what I saw*, *Fifty Years of Denial* and most recently the theatrical movie *The Phenomenon*) and the very knowledgeable Mexican journalist Jaime Maussan. Investigators from UFO groups and a few individual researchers came and went, but the case didn't seem to have the mystery and stimulating aura, or the political theatrics and public appeal of the nearby Roswell crash, so eventually they ignored it again, while Paola continued her investigations.

The transcripts she shared with me (lightly edited here for clarity) leave little doubt about what had happened in the days following the incident.

∧ ∧ ∧

Paola's interview began with Reme's recollection of ***the first day***, as he stated simply: "What happened is that after the crash, we went home, back to the ranch."

Paola followed on: "Can you tell me the date of this? The approximate date of this? We do know it's 1945. It's in the month of August…" (Note)

Reme: And it was, like, the 15th (*This is a contradiction with the date given reported by Jose Padilla, who has a record of August 16th, a Thursday. We've verified that later date*).

Paola: Okay, so whose dad was it that sent you, and you were how old?

Reme: I was age seven and Jose was nine. Jose's dad, Faustino, had asked us, a couple of days earlier, to find a cow that was ready to have a calf.

Note: Pages were being turned quickly: On August 14, 1945, two days before the crash at San Antonito, a team of American scientists had flown to Europe "to collect information and equipment relating to German rocket progress." (Reme Baca, op.cit., p. 98)

Chapter Two - San Antonito, August 16, 1945

Paola: And you were on two separate horses?

Reme: Yes, we were on two separate horses. Here's what he told us: "You know, when you get a chance, I want you to go out and check that cow because it's getting ready to calve, and we want to make sure that we get it before somebody else gets it, and puts their brand on it." What we were doing periodically, is that we'd get on horseback and go up and check all the fences, "riding fence" is what we called it: check the fences, make sure they weren't broken. If they are, you repair them, you have a small tool kit. If a post is down you prop it up, and later on you come back, and replace it. So we'd ride the fences, and when we were done with that, we'd go up to the top of the hills, and take inventory. Jose would look through his binoculars and count the stock. I'd write the numbers down. Count the cows. While twenty-five head of cattle may not sound like much, Faustino had purchased some White Face cattle from Spain, and was in the process of starting a White Face herd. They seemed to do well in that type of terrain. So that's what we'd do. And then of course in the winter when it snowed, we'd sometimes have to break up the ice in the water trough, so that the animals could drink, and transport bales of hay or alfalfa to the windmill area, so the animals could eat.

Paola: This was during the day that you guys went...?

Reme: Yes, this was during the day. Jose would come over on his horse and we saddled mine and we took off. My mom was aware that we're going to work on the Padilla Ranch.

Paola: He was nine and you were seven?

Reme: Yes. We went over, looking for that cow. And so while we were there, it was not abnormal in late summer, to have thunderstorms and lightning and this time was no different, so we took refuge under a ledge. We had dismounted because the terrain was steep and rocky and horses don't do well on rocks, they hurt their hooves. We replaced the bridles with rope and tied them so they could graze, and we continued on foot.

We walked and came up to where there was a clump of mesquite bushes, creosote or *greasewood* as they called it in the day, pine, sage, and cactus. As we walked towards the clump of mesquite we heard a moan and we discovered it was the cow we were looking for, and it had given birth to a calf. This was part of the beginning of the start of a new herd called a "whiteface" herd: a red cow with a

Fig. 8: The first witness was Lt. Colonel William J. Brophy, at Alamogordo AAF base. Credit: Mr. W. Brophy Jr., and *Born on the Edge of Ground Zero*, by R. Baca & J. Padilla, 2011

Fig. 9: Jose Padilla as a child (left) and his great-grandmother (right), Maria Amada Chavez, (1856-1940), local Apache leader.

white face and white feet. Faustino had purchased a cow and a bull from Spain and was breeding them. Whiteface was one of the cattle breeds they used in the United States for meat at that time.

So we found it and then we went down into a little area where there was a ledge. Jose had packed a lunch, a couple of tortillas and I think a couple of apples. We sat down to eat that and the storm and rains came. We got under there so we wouldn't get too wet. Then it just kind of sprinkled a little bit, and was all over. We were getting ready to go up and take another look at the cow and see if it was eating, and take a closer look at the calf. While we were doing that, we heard this loud bang.

Paola (*surprised*): You heard the actual crash?

Reme: We didn't know it was a crash at that time. We heard this sound and the ground shook, and so memories came back of the atomic bomb explosion. *Are they testing again or what?* So we looked around, saw smoke coming from maybe a couple of canyons down, up that way. So Jose says, "Let's go over and take a look, see what's going on." We started walking, and we saw a little smoke coming from that direction. As we reached the ridge, the smoke became intense. Then we worked our way down the ridge, *so we could see what appeared to be a big gouge in the ground*. It looked like a road grader had been in there. We weren't aware that anyone had a 100-foot wide grader, but it sure looked like a 100-foot wide blade, grading about a foot deep. We started walking up this road, it was pretty rough on our feet, and it was warm. *The bottom of our feet felt hot.*

Let's pause here for a moment. What Reme Baca is describing so casually is a remarkable event in itself. For the object to dig such a wide gouge, one foot deep, all the way across the side of the hill, represents a very large amount of energy. In a couple of minutes, the craft carved a wide scar in the landscape that might have taken a couple of hours for a bulldozer. Furthermore, the device must have hit the ground with high velocity and tremendous horizontal momentum. It must have been extremely strong since it didn't break up in spite of that energy release: all things we must keep in mind as we begin to take apart the physical parameters of the case.

This surely wasn't a weather balloon, but it wasn't just a clever model either, based on somebody's idea of a deception, as part of some counter-intelligence operation. The gouge in the earth, the increase in ground temperature, the burning plants, the rising smoke and the fact that the object kept its integrity through the catastrophic event, all point to the fact that what the two boys witnessed was the real thing: a solid vehicle of an unknown type and function.

∧ ∧ ∧

We have strong independent evidence that the crash did happen as the kids describe it. It comes from the crew of a B-25 on a training mission, flying over Walnut Creek, who were directed to the site by Alamogordo controllers when the communication tower was destroyed.

The pilot, Lt. Colonel William J. Brophy, reported seeing the smoke and the bent tower. He circled the area, saw the crashed object in the vegetation and radioed back that "two little Indian boys" were close to the site.

Lt. Colonel Brophy (Fig. 8) told his son in 1978 that he was the first adult witness over the site, but he didn't get near the craft on the ground until the next day, while the two kids were able to approach and examine it shortly after the crash. "There were lots of pieces," according to Brophy.

Paola went back to Reme's recall of the sound:

Paola: (Was it) the same sound as when the bomb went off?
Reme: *Similar* to the sound as when the bomb went off, and it was still fresh in our minds. When the bomb went off, Jose and his mother were up early in the morning. The bomb went off after his dad left for work. Jose's mother looked at the flash thru the crack in the door jamb and as a result of the exposure, she lost sight in that eye. According to Jose, they felt the heat wave, and the rumbling of the ground.
Paola: So the sound was familiar...
Reme: Very familiar. They were closer to the bomb explosion than I was. My bed crashed against the wall and it bounced me out of it; my mother got up and tried to explain that it was probably the storm that was causing all this.
Paola: Going back to the actual story, you heard this sound....And do you remember around what time it was?
Reme: I didn't have a watch. Probably four or five in the afternoon, maybe later.

Chapter Two - San Antonito, August 16, 1945

Paola: I'm asking because you can see what you're looking at, it's not dark.

Reme: No, it's not dark. But as we look up this graded road, *there's a lot of smoke*. So we retreated to where we could get some air and take a drink from the canteen and kind of recollect our thoughts and try to understand what this is all about. I asked Jose, "Is that a plane that went down?" I've only seen planes in the air. We live in a small town. Don't see many planes. Jose says, "Don't know, maybe somebody might be hurt and maybe we need to help them." I said "OK, okay," and so we continued trying to get closer. We could see something over on the edge of that graded gash.

Paola: The path that the grader left...

Reme (*correcting her*): The path that *the craft left*. It doesn't go just straight. It goes, and then it makes a right turn, like an "L". We could see something but you know, there's so much dust in the air, and it's humid from the rain and then some of that brush, *that oily brush is burning*, so the smoke's coming into your eyes, it's really hard to see and make any sense of it at all.

∧ ∧ ∧

Let's pause again here, because it's interesting to hear the boys describing the bushes burning. It has just been raining hard, the air is very humid, yet the creosote is burning, *not the wreckage itself, although it was a natural assumption for Timothy Good to make in his book*. Now the two boys are understandably shaken: They rode out to the property to find a cow and now they are witnesses to a scary, complex event, a strange object falling from the sky. The interview goes on:

Reme: We went back up and rested, returned, and Jose has his binoculars out and starts looking to see what it is. He says, "You know, there's something over there. Let's see if we can get any closer." Again, we try to get closer and finally it starts clearing up a little. The time seems to be going by very fast. We're looking through the binoculars and I could see the hole on the side of this object. *The object is avocado shaped.*

Paola: So it's a round object like an avocado, and you could see there's a hole. How far would you say you guys were from the object?

Reme: I would estimate about a couple of hundred feet.

Paola: And then you saw the inside of the hole from the couple hundred feet?
Reme: No, not the inside of the hole. Jose says, "Look at this!" So I was looking through the binoculars *at these little creatures moving back and forth.*
Paola: Were they moving really fast?
Reme: They were like, *sliding.*
Paola: They were sliding...
Reme: *Not sliding, but more like willing themselves from one place to another.* That type of sliding. *And as I'm looking at that, things began happening to my mind.*
Paola: Oh, really...
Reme: I'm seeing them *and I'm feeling this crazy stuff, like I really feel sorry for them.*
Paola: Um, hm...
Reme: And I really feel sorry, like they're kids, too.
Paola: And you had a concern for them? And you're thinking...did you feel something because of the accident?
Reme: Yes, I think so, I'm hearing this high pitched sound coming from there. We didn't know what to think. The only high-pitched sounds we were familiar with, were of jackrabbits when they were in pain, and also the sound that comes out of a newborn baby when it cries.
Paola: I find this interesting. So you heard this same sound....

^ ^ ^

Until this point, Paola's record describes the two inquisitive boys reacting to what sounds like an accident. They move closer in an effort to help in understandable confusion about what they may find. In the atmosphere of the time, when stories of death and mishaps and aircraft in trouble are common, your main duty should be to help if you're first on the scene, miles away from any support. Reme confirms to Paola that the situation was emotional, but something else then comes into his recollection, something more than the realization that they've just watched an airplane crash:

"*Then we saw these pictures in our heads,*" he says.

Surprised, she asks him to repeat that statement, and he confirms there were pictures, adding: "*I didn't know what the heck they were. I can* remember what they are, *I got pictures*, but I didn't know what they meant then, *and I still don't know*. They must have known we were there."

That statement is important, because it marks the first instance of a kind of interaction between the phenomenon and its witnesses that may be fundamental to our eventual understanding of the UFO problem in general: Witnesses are an integral part of the event.

Paola: Could they see you if they ever looked out? I mean…

Reme: I don't know.

Paola: But I mean, there was a hole…if they looked up, could they see these two little boys?

Reme: Yes, I'm sure they could, if they could see.

Paola: This was about 200 feet…

Reme: Yes, it was about 200 feet from us. However, there was smoke and dust, so it wasn't very clear.

Paola: So what did you guys do, run away?

Reme: We looked at them and now it was starting to get dark, and we had a long hike to get to the horses and back to the ranch. But Jose wants to go in, and I don't.

Paola: He wanted to go inside the ship? Jose wanted to go inside?

Reme: And I'm saying, "Jose, what is it?" His response is "I don't know."

"Okay. If you don't know what the heck this is, *I ain't going into it. There's no way. I wanna go home. I don't want to go in. You'll have to go by yourself.* I'm going home,

I'll meet you at the ranch."

And he says, "Well, let's watch for a little while; you know, maybe you're right. I don't know what they are. They kinda look like kids, *very strange kids*."

Jose has seen the beings and now he's confused. He realizes that this may be too much for seven-year-old Reme, and he defers to him, although left to himself, his curiosity would probably drive him straight to the craft. "So you had a whole conversation about this?" asked Paola, and Reme recalls: "Oh, yes!" And so Jose says, "Well okay, let's just watch for a little longer, and then we need to get back home. Your mom's probably worried, it's getting late, and I'm sure dad's worried…" The reality of home and family reasserts itself and becomes the rationale for leaving the scene, a critical decision to which we will return later, when Paola speaks to Jose himself and gets his side of the story. But the kids' duty to their parents wins in the end, and they go home.

Paola: Can I ask if you and Jose had a conversation about what you saw, in those years?
Reme: Jose left San Antonio in 1954 and I left in 1955. During the years we were there, yes we talked about it. *From 1955 to 2002, we had no contact.* Since 2002, we have compared notes.
Paola: You have compared notes? And does he remember things?
Reme: Better than I do. *He has a photographic mind.* He started school at the age of four.

Then it was the end of the first day, the day of the crash. They had not attempted to go inside the object at that point. Instead, the two boys agreed it was wisest to go home "because it was getting late."

So they walked down the slope and got on the horses, and started off.

"It was getting dark then, and it was pitch dark by the time we got to the ranch," Reme recalled. "And Jose's dad was waiting for us, he was worried. So we went in, and Jose told him the story about the cow, and then he started telling him about the crash; and I told him what I saw, and so his dad says, 'Well, the first thing we got to do is we got to get you home. We'll look into this in the next day or so. It probably belongs to the government, and that's probably it. We need to, maybe, stay away from there.' And so they drove me home, I left my horse there and they took care of it. They drove me home, and Faustino had a long talk with my mom regarding the object that we'd discovered on the Padilla ranch. Faustino emphasized it might endanger his job, since my dad worked for the government. And so that was basically what happened that night."

At this point on the tape, there is a further exchange between Paola and Reme about the size of the object, which the two kids actually measured by "stepping" it when they came back to the site in the following days. Jose Padilla, who paced the ovoid craft as best he could without being detected, said it was about 25 to 30 feet long and 14 feet high, which he knew "because of the rafters of a house are 14 feet tall."

Not bad thinking for a nine-year-old kid.

∧ ∧ ∧

CHAPTER THREE

THE PADILLA RANCH, AUGUST 16-20, 1945

It was now Day Three, Saturday August 18th and two men, namely Jose's father and a State policeman, were about to climb aboard the crashed vehicle from Elsewhere. Nothing much had happened on Friday, because the kids had been kept away by Jose's father, Faustino, who needed help with gardening chores, so Jose and Reme only drove back to the site on Saturday, using the family truck, guiding Faustino and the officer to the actual spot.

At that time, namely Day number Three, as Reme Baca told Paola when she continued the interview, "Jose came over to my house, and I went with him to his house, where we met Eddie Apodaca who was a State Policeman, and a friend of the family. Faustino had asked him to go with us to the crash site. They rode in the State Police car, and we rode in the pickup truck. We drove as far as we could get with the vehicles, and we walked the rest of the way to the crash site. When we got close to the crash site, looking down from the hill, we couldn't see the object."

Paola: What do you mean? Did you get very close to the crash site?
Reme: We're not talking about flat land here. We're talking about hills, canyons, and arroyos.

Standing on top of a hill, looking down towards where we'd seen the object, it was no longer visible to us, at that time. No explanation why. We simply couldn't see it. It seemed gone. Jose says "Well, I don't know what's going on here." Eddy and Faustino said, "What did you say you saw?" My response was, "It's down there, but we can't see it." Faustino said "Let's walk down there and take a look." We started walking down, and then we saw it. The object had a lot of debris over it and so I'm asking Faustino, "How come we couldn't see it from up there?" His response was that he didn't know.

Paola: You're saying it was almost invisible?
Reme: I almost couldn't see it. Then we got there and they said "Okay, you guys stay here and we're going to go in."

Reme pauses here, as if puzzled by his own recollection, and it is Paola who prompts him:

Paola: So, Reme, they went in… What did they find?
Reme: Whatever they found, they didn't tell us. What I do know, is they found a complete change of attitude. When we were coming down the hill towards the crashed object, they were doubting us a lot.
Paola: Yes, I know, I know.
Reme: So they went in and we stood there, sat down and watched them. And they were in there five or ten minutes and came out. *They had a change of attitude, a complete change of attitude. They were almost like different people. They had seen something they'd never seen before.*

What did the grown-ups see at that point, that made such a powerful impact on them? Whatever it was, the intensity of the situation had escalated, along with the need for absolute secrecy. On Paola's recording, Reme goes on.

Reme: They came out and said, "Okay. Here's the way it is. I want you guys to listen. This is very difficult. You're under oath. You don't tell anybody about this, not your brother, not your cousin, not your mother, not your father, that's our business. We'll take care of that. And the reason for this is, that you can get in trouble. We want to keep you out of trouble." So we agreed to that and they gave us a really big lecture, and so we took it very serious.

Paola: But did they ever tell you what they saw inside? (Note)
Reme: No.
Paola: They never described it?
Reme: No. They didn't say what they saw.
Paola: They didn't... But they didn't see any of the creatures, because they weren't there?
Reme: They weren't concerned. Because we asked them about the creatures: "Where are they, because we can't see them through that big hole? There's no creatures there." They said, "Well, you know, maybe they took off. Maybe somebody took them. Maybe...."
Paola: Was there any evidence the Army had been there, any?
Reme: Evidence? We saw something like a broom (mark), or rake mark, but then again, it could be some animal, insect or snake that made those marks.
Paola: Because logically, if the military had taken the creatures, they would also have had to show that they had been there in some way. In other words, they waited at least 24 hours before taking the craft?
Reme: Maybe they did show that they had been there, but we were not aware.
Paola: Well, before taking the craft?
Reme: No, the craft itself took days to get taken out of there!
Paola: How many days?
Reme: Oh, several days. First, they would bring in some road building equipment, build a new road and a gate, bring in a semi-truck with a low-boy trailer, build a frame on the trailer, bring in a crane, and load the craft onto the tractor trailer.

∧ ∧ ∧

Now the kids are back home in mid-afternoon, but they are eager to return to the site. It takes a long time to make the trip on the highways, but it can be done in half an hour on horseback, cutting through the hills. This time, no adults are with them. And they have ready excuses, in case questions are raised.

Note: In *Born on the Edge of Ground Zero*, Reme Baca writes: "While Apodaca and Faustino were in the object, *they were always visible to us*."(**12**)

Paola: Okay. You were going to go back in there…

Reme: We were working in that area: We checked that fence, too. We had some fences to fix, and fence poles to replace. There were cattle with calves around there, also.

Paola: So what happened?

Reme: Finally, we got there, we were on horseback, and came in from a different direction looking from the opposite side of the ridge, *we saw some military people picking up stuff.*

Paola: Okay. Well, that's what I had just asked you before. How did you know the military was there before, you said the creatures weren't there…

Reme: The military wasn't there all the time.

Paola: But the creatures were gone, and I was wondering, the military must have been there to take them?

Reme: We didn't see the military take them. If they did, it was before we arrived. But we never got to check the craft, all we got to do was go down and get some of the debris and threw it in this crevice and we tried to cover it with dirt and rocks. After the two Jeeps left, it was already getting dark and we had to get home.

We should note that Mr. Padilla, when he was interviewed by us, only recalled one Jeep at this point, not two. It seems that the Army detail sent to recover the craft came from White Sands Proving Grounds to the East (see map on Figure 14). The military contingent would have driven west on Old US-85, but the kids couldn't have known this at the time.

The conversation continued about the discarded material picked up from the object. "What did that material feel like, the material that you threw into the trench?" She asked. Reme answered: "Kind of like this piece that I'm holding in my hand." (Note: A sheet of metallic foil.)

Paola (*holding it*): It was like this?

Reme: It was hard. On the first day, I had gotten a piece of that aluminum foil type, and showed it to Jose. It reminded me of the aluminum foil that came in the Philip Morris cigarettes that my mother smoked. I took that and put it in my pocket…

Paola: Whatever happened to that?

Reme: I used it to repair the windmill cylinder.

Chapter Three - The Padilla Ranch, August 16-20, 1945

In a separate conversation, Reme later described that piece of foil to Paola: *"It was about four inches wide, and it was almost fifteen inches long."*

Everything can be put to use on a farm, sooner or later: the kids used the metal from the unknown craft to fix the backup pin of the well on Faustino's property, and they never thought about it again for over fifty years, until the ancestry search reunited them.

∧ ∧ ∧

On Paola's tape the interview continues, centered on the events on the fourth day, Sunday, August 19th. Reme recalls that Jose came to his house: "We picked some chiles, green peppers, tomatoes because we had a vegetable garden and they didn't, and we filled a couple of bags with vegetables, and we took one to his house.

We went in the back door. And as we go in, there was a military vehicle in front and there's a soldier there at the screen door talking to his dad, so we go around the back and in through the kitchen to join them. Faustino says, "Come on in here, boys." So we joined him and he's talking to a Latino, Sergeant R. Avila, and he invites him in. Sergeant Avila says, "I'm with the US Army and what I need to do is get permission from you to go in and cut the fence and put in a gate, because we have one of our experimental weather balloons that inadvertently fell on your property."

Paola (*laughing*): He called it a weather balloon? Those words?

Reme: An *experimental* weather balloon! And so, "we need to recover that, so we need permission to do that." So his dad says, "Why can't you come in through the cattle guard like everybody else, instead of cutting my fence down?" And Avila says, "because the equipment that we're going to bring in is wider than your cattle guard, it won't fit through there." He says, "in the meantime, you have a gate that locks up and we need to have a key so we can get in there and cut that fence, and put in a gate." He goes on, "We'll put in a good gate for you. And then we need to bring in some road-building equipment, some graders and so forth and see if we can grade a road to get that truck in there to get that weather balloon out of there."

So finally Jose's father says, "Okay." They both spoke mostly in Spanish. He says, "Okay, go ahead and do that." And Avila says,

"Keep an eye on the place and make sure nobody goes there because you know, this is really important, you know, we don't let anybody know about it. We don't want to cause any trouble for anybody, and so try and keep an eye on it, so nobody that hasn't any business going there, doesn't go there."

And so, Faustino said, "OK," and Sergeant Avila left, and that's when they officially began the process of preparing the area to take the object away.

From Jose's testimony, then, construction of the new gate and the new road could have begun on Monday, August 20th, although it would have taken a couple of days to build a dirt road strong enough to safely hold an eighteen-wheeler hauling away an "experimental weather balloon" that weighs several tons...

∧ ∧ ∧

Reme told Paola that "the recovery wasn't like what we now see in UFO movies, people in purple uniforms dropping in from helicopters, everything sanitized. Nothing like that!"

Paola was surprised: They weren't wearing protective clothing…?

Reme: They were wearing fatigues, they put up a tent, played a radio, western music.
Paola: You were watching them, then.
Reme: Yes, we were watching them, as often as we could, sometimes in the morning, and evening. It was our job to check and maintain the fences, keep track of the herd, including horses. We could hear the radio music going. There was one guy there at the tent, and two or three working, picking up the debris.

They bring in this tractor-trailer, they have a welder, acetylene welder, and they build this rack so they can get the craft on it because it's got to go on sideways. Then we figured out they were doing that, because they had to go under the overpass at a forty-five degree angle in order to clear it.
Paola: Did they tie it up or put a tarp over it?
Reme: Yes, put a tarp on it.
Paola: And tied it up….

Chapter Three - The Padilla Ranch, August 16-20, 1945

Reme: These soldiers were kids, and they went to the Owl Bar and Café a lot.

Paola: Was that the Owl...?

Reme: The Owl Bar and Café. And so the Owl Bar and Café was run by Estanislado Miera. In the parking lot, they had a basketball hoop, where we played. They had what they called a "Fountain," a soda fountain, and they sold ice cream and shakes, food. They also had a jukebox. So that's where the soldiers went to socialize. And so we would go there and play hoops and then sometimes Estanislado would come out and ask us to help him. Sometimes we would help grind up meat for hamburgers, wash dishes, clean the place up.

Paola: And so these guys went there?

Reme: And yes, that's where they went for lunch, that's where they went for dinner.

Paola: And you saw them pick up debris at the crash site?

Reme: Yes...

For me, reading this in the twenty-first century, Reme's transcript seems almost too simple, too matter-of-fact. An extraordinary object drops from the sky on that isolated slope; the military is unprepared to deal with it, so they improvise: they send a group of young recruits who don't care and over a few days they clean up the site, pick up what they can of the strange material scattered around by the shock, and they bring equipment to evacuate the object as soon as possible.

None of the precautions we will witness in later UFO recovery efforts or in Spielberg movies, with sanitized equipment and crews in special biohazard suits, are in evidence in San Antonio in 1945. Nobody even thinks about it. The soldiers come and go, they eat at the nearby diner while the kids, equally unsupervised, pick up stuff on the hillside and save pieces to play with them later: There is some bright spider web, or angel hair, which we glorify now by calling it fiber optics: It is saved as a curiosity. And the metal strip with the strange properties is put to good use in later years when the family needs to fix a windmill.

But it all makes sense

Of course it would happen this way: nobody could have been prepared for such an event. Reme and Jose act under the impulse of the moment, watching everything, waiting for an opportunity to grab some piece of evidence, some object that might be useful, or a toy, or a trinket.

The opportunity arises when the recovery comes to its conclusion. Using their brand new road, the Army has driven up with a large truck and they've brought a crane which, knowing the size of the object, must have been 15 or 20 feet tall. On the tape, Reme tells Paola that on the final day, he saw them lift the craft and drag it onto the tractor-trailer.

Paola: Did they ever see you?
Reme: I don't know if they ever did, or cared.
Paola: In other words, you were like part of the scenery...
Reme: Well, you know, they weren't looking for us, and there was vegetation on the side of the hills and we weren't very tall, so it was easy for us to hide.
Paola: But you didn't go and talk to them or anything?
Reme: Oh, we would sometimes talk to them at the café, but not much, because we didn't have anything in common. *The work they were doing didn't seem all that important to them. It didn't seem to be a great deal to them.* We don't believe anyone was aware of how important this object might have been, certainly not us. Years later, one of the soldiers married Jose's cousin.
Paola: You just said one of them married Jose's cousin, and the obvious question everybody would have... Did this military man who married Jose's cousin ever talk about this incident?
Reme: With Jose? No. With Jose's dad.
Paola: With Jose's dad? This military guy did. Do you remember what was said?
Reme: I wasn't there. But Jose would know. It's my understanding that throughout the years, he became more unconvinced that it was a weather balloon... He was just doing his job, picking up the debris, looking forward to completing his assignment and going home. The war had ended, and a lot of the soldiers had been restricted to *Stallion Site* for the last 90 days.
Paola: He didn't know? So, his job was just to do the recovery. But he thought it wasn't a weather balloon...So, you guys were at the soda fountain, and then what happens ...?
Reme: We'd go there, buy a coke, and listen to the music...

Chapter Three - The Padilla Ranch, August 16-20, 1945

Of course they would. The famous Roswell crash and the Kenneth Arnold sighting, with all the excitement in the American press, were still two years away in the future. The term "flying saucer" hadn't become part of the vocabulary and anyway, that craft wasn't a disk. No media were covering it. No report had been made except to White Sands headquarters. It was just a strange private adventure for Jose and Reme, caught in the middle of a very focused military operation that was complex, unique, secret, and evidently improvised.

Yet the kids did much more than listening to Bing Crosby or the Andrew Sisters at the café singing "Don't fence me in," or Doris Day in *Sentimental Journey*...

∧ ∧ ∧

After those days of work at the site, the craft was safely leaning on its side so it wouldn't hit the overpass, covered with a light blue tarp tied to the sides of the tractor-trailer rig. The soldiers were ready to drive it away, back to the White Sands Proving Grounds.

The Army detachment would have left the scene no sooner than Thursday, August 23rd or (more likely) Saturday, August 25th. (Note)

Paola asks about the scene on that last day. Reme answers: "It seemed that the guys weren't even aware that we existed. They were predictable. We pretty well had it figured out, and on that last day, Jose came and got me and we went to the site, sitting in the brush where they can't see us. We watch them drive the truck outside the gate and they got the tarp tied up nice and neat. Jose said, "I think they're going to take it tonight." I said *"Yeah, how about a souvenir?"*

Note: As mentioned above, the son of pilot Bill Brophy has reported helpfully: "My Dad's B-25 bomber group, the 231st at Alamogordo Army Air Force in New Mexico, recovered this spacecraft August 15-17, 1945."

As we've seen, the object must have stayed at the site much longer, although indeed Brophy, Sr. may have been able to secure the site on August 17th if his commanding officer, General Maurice Arthur Preston, did leave him in charge there until the next day when Army Colonel Turner took over. But the detachment seen by Reme and Jose doesn't appear to be a team of aviators, and the object could not have been physically evacuated until a week later.

Paola (*surprised*): You said that?

Reme: Yes. During the war we lost so many relatives, that it was not unusual to have something to remember them by, when we said our prayers. Because when they died in the war they never came back. They brought a lead coffin and a couple of guards with it, and they buried them.

Jose says, "Let's head down and wait a little while until they leave, and then we'll go."

We waited for a while and then everybody took off. They had these military pickups and they took off. So we know where they're going, they'd be gone for a while. We worked our way up there and where the crevice was, they had run the grader through it, so nobody would even know that a crevice existed. Then we worked our way outside the fence, towards the back of the truck and stepped it off. If you made a big enough step it was three feet. Maybe we were off a few feet, but that's the measurements we had: 25 to 30 feet long and about 14 feet tall. And then we looked at the underneath part of the craft, because we had not seen this part of it, it was partly underground. So now we get to see the whole thing. *Boy, this thing is a monster, it's big*. Now we can see the bottom. And in the bottom it's got like three little indentations, little grooves under there, on each side. (…) And so, Jose pulls part of the tarp off, exposing the gash on the side of the craft, while I hold the tarp open. Jose climbs into the gash.

Paola: He went inside the hole?

Reme: Yes, and I was partially in, holding the tarp, letting the light in. First, there's nothing hardly in there.

Paola: But he could see the shape of it? Like, if there was any rooms? Was it smooth all the way around? If there were panels, if there were… try to explain it to me!

Reme: Jose said there were… like ridges every so many feet.

Paola: Did he see any panels, like control panels?

Reme: No. He did see like a panel, like this big of a panel. We're talking, maybe about two and one half feet panel.

Paola: Was it attached to the wall, this panel?

Reme: To the bulkhead, the rear wall, maybe.

Paola: This was the panel which is inside, as you said, *bulkhead*. But it's against the wall, this panel. No? It's attached to the bulkhead?

Chapter Three - The Padilla Ranch, August 16-20, 1945

Reme: Yes, to the bulkhead. What would be the rear wall to us.

Paola: So how fast could he pull that thing off? I mean, did he pull it off?

Reme: He tried to jerk it off and he couldn't, so then we went and got a cheater bar from the front of the tractor-trailer…Something like a crow bar, it's called a "cheater bar" in the trucking industry, it's used for testing the tightness of the chains holding the load down.

Paola: You described the pins and what they were like…

Reme: Yes, a one way fastener. They go in one way and they can't come out. They were serrated fasteners that were inside the holes, and that's what was holding this bracket-type piece on the panel that was located on the bulkhead (rear wall). The pins were yellow…

Paola: The pins were yellow. Did you see any other colored stuff in there?

Reme: Silvery colored strands of what I would compare to angel hair. No seats or anything, nothing…it must have been cleaned out, or maybe there weren't any. Couldn't see any instruments, like gages, clocks, steering wheel, brake pedals, nothing like that.

Paola: Was it gray inside?

Reme: Part of that ship was darker on the bottom part than the top, lighter gray.

Paola: But did he, like, race out of there? Because he thought they'd discover him? Or were you guys relaxed because you knew where the military were and they were going to take their time, so you just took your time?

Reme: We tried to hurry, we were afraid of being discovered. Relax? You gotta be kidding!

I haven't relaxed since then!

Paola: This is pretty heavy metal, though….not really, no? Did it feel like an earthly metal? You couldn't tell! But the piece like the Phillip Morris package was different. Where did that come from?

Reme: When it first crashed and we first went in to the crash site, there were some pieces of material that looked like angel hair. It was used in that era, when people didn't have electricity to decorate their Christmas trees. The material was similar to angel hair. I also found a piece of shiny metal..

Paola: Were there I-beams?

Reme: I would not have known what an I-beam looked like.

Paola: You didn't see any structural beams?

Reme: Not that I remember.
Paola: So you wanted a piece of the metal…
Reme: …and so we took that.

^ ^ ^

With that part of the discovery clarified, one major question remained: What happened to the creatures the two boys had seen after the crash? What did they look like? Were they human? And if they were not human, what could we learn about them?

Paola moved the discussion back to the beginning, the very first afternoon during the storm, when the object plowed into the hillside and came to rest with a big open gash, and several creatures in obvious distress inside the damaged craft.

Paola: Did you both ever, when you were discussing as kids… Did you ever discuss the beings?
Reme: Yes. Their heads were comparable to a "campamocha." That's what we saw.
Paola: What do you mean, you saw them and they looked like a bug?
Reme: Yes, they were ugly to us at first. Their heads looked like a campamocha.
Paola: Would you say it in English?
Reme: The closest translation would be like… heck, a bug, *praying mantis*.
Paola: Oh, that would have been…
Reme: Big, bulgy eyes…you know. Everybody calls them "Greys," I guess, but I haven't seen a Grey, so I wouldn't know.
Paola: But these could have been a totally other thing…
Reme: They had big bulgy eyes, we don't know whether they were exactly four feet tall, it's just an estimate. Four foot tall, and they were real thin, needle-thin arms.
Paola: I don't know about "needle thin." And…
Reme: I don't know how many fingers.
Paola: But I mean, they seemed to glide. Were they wearing outfits or…?
Reme: Well, either they were wearing real tight coveralls, or their skin was real tight.
Paola: What color, still grey, the coveralls?
Reme: Yes, light grey.
Paola: And the head was pretty big? I mean, proportionally?

Chapter Three - The Padilla Ranch, August 16-20, 1945

Reme (*nervous*): The head seemed pretty big, and it was similar to a campamocha.
Paola: Okay, that's okay...
Reme: Not like they say...
Paola: No, no, I understand. No, no, no, because *I've heard this before. It's okay.*
Reme: If you see one, I'll have to get a picture of one, but *campamocha* well describes it.
Paola: And you said they slid, instead of walking or running. They "seemed" to slide.
Reme: It seems like they did. *Like they willed themselves from one place to another.*
Paola: But you knew something, they must have connected with you at some point. Well, you said you had images coming in your head.
Reme: Yes, I'm sure.
Paola: You know, I wouldn't see eyes unless eyes were looking at me.
Reme: Jose and I were looking at the craft through one set of binoculars, we were taking turns. He was looking, but we couldn't directly look into their eyes, that I can remember, it's pretty far, I know, but *what we felt was this pure sorrow, really felt sorry for them* because we could feel their pain. They seemed like us, children.
Paola: Oh, okay. That was certainly interesting.
Reme: I have no words for that, to compare something like that! They seemed like they were hurt.
Paola: They were hurt. (...)

Hearing again that part of Paola's dialogue with our first witness, which I have lightly edited for clarity and partly summarized here, I believe it highlights several remarkable facts, not the least of which is Reme's strong emotion as he recalls it so vividly, seventy years later, that Paola has to calm him down: "That's okay..." she says.

But it's not okay. And we will see in later chapters that it isn't okay for Jose, either. And it wasn't okay for Faustino and the police officer when they first went inside, and came out of the craft overwhelmed with a strong emotion. All that is real: we're not dealing with a simple observation of some damaged vehicle.

Nor is this a case of reconstructing a scene days later, as in Roswell: We have two witnesses on site during the crash, and two adults visiting the craft and stepping inside it less than forty-eight hours later.

In an extraordinary description of his own emotions, Reme recalls that the anguish of the little creatures seemed to get transmitted to him. When the craft had passed over the landscape and plowed into the hillside, triggering a rise in temperature, fog and smoke, time itself had seemed to change. Now the kids "feel pure sorrow," as if that overwhelming emotion had been communicated to them by these creatures that are not human, creatures that can move from one place to another seemingly by instant gliding, yet "seem like us, children."

Paola: Anything else that you and Jose... Did you talk about them at all?
Reme: Yes, we talked about them, when we were sure no one else was around.
Paola: What did Jose say about the beings? I know how you felt about them. What did he say about them? Did he feel like you?
Reme: The same, yes. Did he feel sorry for them? Not as much as I did, but he did.
Paola: Were you terrified when you looked at them, or did you want to get closer, or did they just disgust you, or you just felt sorry for them, or....?
Reme: Normally, I would feel sorry for friends, relatives if something happened to them. But I didn't know these creatures. We were curious. They were strangers, we didn't know who they were, but we knew they were different.
Paola: Oh, okay. So you felt their emotion.
Reme: That's right.
Paola (*as the implications sink in*): Oh, my...You felt their emotion...
Reme: And so those sounds, we tried to figure out what the sounds were. We attributed them to coming from them.
Paola: That's probably where they were coming from. How long do you think that experience lasted, when you were standing there doing that?
Reme: All the time they were there.
Paola: Which was?
Reme: Probably a half-hour to 45 minutes.
Paola (*stunned*): You stayed a half hour to 45 minutes where those beings were? You weren't scared?
Reme: We were scared, yeah.

Chapter Three - The Padilla Ranch, August 16-20, 1945

Paola: And you still stayed a …

Reme: We still stayed there. Jose was curious about the creatures too, he wanted to help them. Yeah, Jose tried to talk me into going into the object to help them, and I'm trying to avoid it, yet I'm concerned too.

Paola: Jose was going to go and...

Reme: We don't know what the heck they are, who they are, and what they are doing there. I don't feel good about it.

Paola: You're not quite sure what that experience produced…Well, I can see that you could be confused about…that's a long time. I mean, I think regular people would get scared and run away after they saw them, but you stayed and….

Reme: *Something kept us there.*

Paola: Something kept you there. Because you were trying to figure it out.

Reme: Yes, trying to figure it out. So then eventually we had to leave, to go back to the ranch.

It is interesting that Paola, always the perfect, impartial interviewer, finds herself compelled here to advance her own, rational explanation for the kids' paralyzing fascination at the scene: "You were trying to figure it out," she suggests, offering a conventional, reassuring rationale for their confusion. And we see Reme, who was clearly emotional a few minutes before, jumping at that chance to reconcile that long time he and Jose spent on the spot staring at the hurt, crying creatures with what a perfectly normal person would do if confronted with any strange scene: "Yes," he says, "trying to figure it out."

Which is what we're still trying to do here, seventy-six years later.

^ ^ ^

CHAPTER FOUR

THE SECRETS ARE KEPT (1945-2003)

As they grew up, Jose and Reme, best friends at school, occasionally spoke of their experience in private, but they were smart enough never to talk to others, even as prominent UFO sightings made the worldwide news, like the Kenneth Arnold case of June 24, 1947, and especially the Roswell crash of early July 1947, some 120 miles due East of Trinity, or the Socorro landing in April 1964, only some eight miles north of there.

The expression "Flying saucer" eventually entered the English language, yet the two boys, still wary of retaliation from authorities, kept their awesome secret from friends, journalists, and prying ufologists. Furthermore, they hadn't seen anything remotely resembling a "saucer."

The Army's attitude, rigid in extreme secrecy and in absurd "those were weather balloons" explanations, and the lack of interest from scientists, combined to create an atmosphere where any clever witness understood the wisdom of shutting up. That is still true today, in Europe and the Americas. Many of the best-documented, best-observed cases don't get reported, except, occasionally, to a few trusted individual scientists known for having taken the risk of exposure.

On April 24, 1964, Socorro police officer Lonnie Zamora spotted a landed UFO in a gully, due north of the Padilla Ranch. By then, both boys had grown up and taken up jobs outside the area. While still in high school, Reme had moved to Tacoma, Washington State.

Jose Padilla had returned from Korea and had become a State Trooper in Rowland Heights, California. They were busy working, raising families. They remained uninvolved in the raging media controversies about UFOs.

Scientists and the public in general may not be aware that many witnesses do the same today, even if they have in their possession what they think might be evidence of the reality of an unknown phenomenon; or, I should say, *especially* if they had such evidence in the form of a souvenir, whose actual ownership might involve unknown entities. Jose and Reme didn't actually own what they called the "Treasure," although it was found on family land, Indian land, but the Army didn't have any clear right to it either (Note), and the original owners, whoever they were, Aliens or not, hadn't made any renewed effort, physical or legal, to claim it back; so the best course was simply to shut up, as the two men later recalled in their conversations with Paola, who sought to reconstruct the chain of custody over this gap of many years.

When she asked, "So you got the piece, and then who kept it?" Reme's answer was: "Jose kept it for probably a couple of days, and then, after that, he brought it to me and I hid it under the floorboards at the storage place across the street. Jose told me that some soldiers had contacted his dad, and wanted permission to look thru his tool shed and his house, and he didn't want to get his dad into any trouble."

Paola: And then, what?
Reme: The military went through the shed and his house. They took metal, old weather balloons that had collapsed on the ranch and voter registration material he'd stored there. Then the sheepherder, a longtime friend of my dad's, came into town herding the sheep to the stockyards, where they were kept overnight and loaded into railroad cars the next day and shipped out. In fact, we went with him over to the stockyards, where they camped out--they used to make such good soup! And we joined them for dinner, then came home. On the following day, the sheepherder moved in the storage house, and gave my dad a young lamb.

When Jose and I pulled that piece off that craft, our souvenir, we had named it "Tesoro."

We were the only ones that knew its name. Translated, it means "Treasure".

Note: The Army would have been in the very uncomfortable position of explaining to a federal judge what a spinning metal bracket was doing inside an "experimental weather balloon" weighing several tons, whose origin they couldn't explain.

Chapter Four - The Secrets are Kept

Paola: Okay. Treasure.

Reme: So that was our Tesoro. The sheepherder comes over to the house one morning while we're just finishing breakfast. My dad's home on vacation, and he's not aware of our tightly held secret. The sheepherder knocks on the door, I answer it and he says, "Can I talk to your dad?" "Sure, come on in." Dad says, "Come on in, Pedro, let's have a cup of coffee; we're just finishing up our breakfast". So, we're sitting, finishing up and so Pedro says, "Alejandro, you know, I'm going to have to leave this place."

"Why?"

"Well, he says, you know, last night I was asleep and I got woke up. I saw this light out there by the well. There was this light out there and I...."(*Reme hesitates, emotional*)

Paola: Okay. He saw a light by the well...

Reme: ...by the well, and he says, "I looked out the window and the next thing, there's these three critters in my room, and the door's locked..." And so he pointed towards the floor: "... *and they're saying "Tesoro."*

Paola: Oh, no! You never told me this part of the story! This is incredible! Oh, my Lord....

Reme: And so, they're pointing there. And so he says, "*Tesoro, there's a treasure down there.*"

And so these guys are doing that, Pedro says, "I got my rifle and I'm going to shoot them because they have no business in my house. And so I got my rifle and they're gone. But you know what, they went right through the wall. Can you believe that, Alejandro?"

My heart is pounding, and I am silently praying, I don't want to get in trouble. And my dad doesn't know anything about the Padilla ranch experience (*with the bracket – JV*).

Paola: Okay.

Reme: So my dad says, "all right." He says to my older brother, "Dave, let's go over and check, bring a shovel and a crowbar."

And so he gets the crowbar and undoes the floorboards and he steps down and says, "Where?"....And Pedro pointed "Right there. Right in the center of the room."

I'm silently praying, "Oh, God, I hope they don't find it."So he digs in the center, he can't find anything, and he digs around with the shovel, and there's nothing. He says, "There's nothing here," so they nail the boards back on, and then my dad says, "Well, it'll probably never happen again. Don't worry about it. If it does, let me know and we'll check it out again." So everybody was happy, and that was the end of that. I saw Jose the following day at the Post Office, and I says, "Hey, you need to come and get that Tesoro, because there's too much happening around."

So he comes over and gets the Tesoro and takes it home, and puts it with some other stuff underneath his house. At that time we had space under the buildings, due to floods.

So Jose puts the Tesoro in some boxes underneath his house and that's where it lay until 1963 when he went back after he had moved to California.

Jose had moved to California in late 1954. In 1963, he went back (to New Mexico) to repair his windmill since he'd purchased some used windmill parts. The caretaker drank a lot, and Jose had a hard time finding him. Jose decided to take all the boxes home (back) to California, and put them in the attic in his garage.

^ ^ ^

Our two witnesses wouldn't communicate again until an Internet ancestry search, of all things, put them in contact again through the network, and their interest in the incident was rekindled.

As Reme would tell Paola Harris in 2010, "Most of the contents of the boxes were old dishes, bottles, odd papers, letters, magazines and useless junk, and that's where it all remained until 2001-2002 when I met his son on the Internet, and his son informed me that his dad's name was Jose and was from San Antonio. I called him and rekindled our youthful experiences and the discovery of an object that was shaped like an avocado, that had crashed on the ranch when we were little kids looking for a cow that was ready to have a calf.

"What the heck did we call that piece that we took off that object? Del Oro? Socorro? Ah, Tesoro! Yes, that's it! *Tesoro*. You know what," Jose says, "I bet you it's still there, way back in the attic, it's been so long that I had forgotten about it. Let me take a look and see if I can find it."

English author and noted UFO analyst Timothy Good was given a piece from the object for testing when he met with Reme in Seattle. In his book *Need to Know*, he recalls examining and analyzing the "souvenir:"

Chapter Four - The Secrets are Kept

It appears decidedly terrestrial, a point conceded by the witnesses. Looking like a bracket of some kind, it is 12 inches long, weighs 15 ounces and contains a number of holes for fasteners. A section cutoff for analysis, as well as acid tests, reveal the metal to be 200-series aluminum, or similar. A smaller, semicircular piece, believed by the witnesses to have come from the craft, was found years later at the ranch entrance. Tests indicate 330/380 series aluminum, or similar. **(13)**

When she interviewed Reme and his wife Virginia ("Ginny") in Gig Harbor some three years later, Paola carefully reconstructed the history of the object and the chain of custody, to get the story straight.

Paola: This is the same piece that you remember, that very same piece?
Reme: Yes. *Tesoro*. He found it, and so he Fed-Exed it.
Paola: He Fed-Exed it to you because you wanted it, right?
Reme: Yes. Because I wanted to get it tested.
Paola: How did Jose and you, Reme, get together to talk about your experiences?
Reme: Well, I wanted to begin to research it. Find out...
Paola: Okay, so 1994, ten years, more than ten years ago...
Virginia: He had to find Jose. After he found Jose, then everything started coming out.
Paola: When did you find Jose?
Reme: You know, what my problem was, was trying to recollect from way back when...
Paola: But what year do you think you found Jose?
Reme: In 2002. It was after his surgery. He had open-heart surgery. So I was doing everything. In fact I took a trip to New Mexico with some guys from California and...
Virginia: We lived in California.
Reme: Yes, we lived in California. It was in July, I think... One of the first projects after retiring was to do our genealogy, and I was using the Internet, and by chance I met a person by the last name of Padilla and we started talking back and forth, and I asked him who his dad was, and he said "Jose" and that he was born in San Antonio, and I informed him that we were friends.
Paola: How big was San Antonio?

Reme: Well, they had six original families. The population was not that big, between 50 and 75 people in the area.

Paola: So at the same time, do you remember the year this happened?

Reme: It happened around late 2001 or early 2002.

Paola: So it's 2002, after all of these years, you were able to connect with the other little boy who was on that hill. You were seven and he was nine, and you were able to connect and compare notes?

Reme: Yes.

Paola: And you said that he said something very significant, when you called him. You said...

Reme: He says, "Did you mention what we discovered to any members of your family? And if you did, what was their reaction?" And I said, "They didn't believe me."

Paola: Okay.

Reme: And he says, he had the same problem.

Paola: But you were telling me that for the longest time your wife, Ginny, didn't believe you until you had a sighting, and then everything turned around.

Reme: That's right.

Paola: And then she's become a very good partner in this search.

Reme: Very supportive, yes.

Paola: Very supportive of it. Okay, so very quickly, you've made a model of the craft that you and Jose saw, and this is the hole that you said...

Reme: Yes. That's a replica of the craft and the hole that...Essentially, in reenacting what we called the fly-by, that came by a tower and it was either radio or radar tower about 50 feet tall, and that tower had what they called at that time a wind charger. And so in talking to other engineers and so forth, this craft may have been traveling by there and may have been affected by lightning hitting the tower. Since the tower would have been grounded, the juice goes into the ground. If the craft was not grounded, and it got caught in between it, some part of it might get fried.

Paola: Okay, and then later you got other pieces of metal which you took in the space of like a week that they did the clean-up, but (there's) a piece of metal that you just looked at recently, that has a lot of little bubbles on it that looks like it's been subjected to heat.

Virginia: I believe they call that EMT heat. And that's a very high, intense heat that melts, that has to be really hot to melt that.

Paola: So what makes us think that this piece which has ridges and little circles, what makes us think that this is an ET piece, or extra-terrestrial? You've had it analyzed….

Reme: That particular piece has a little carbon type of hairs in it and so if you notice that piece, even though it has a few melts around there, it transfers that heat from one side to the other. And so that prevents meltdown.

Paola: Okay, and you have had this analyzed, and you have all the analyses of this saved?

Reme: Yes.

^ ^ ^

Coming back to the samples obtained by Reme and Jose, Paola went on: "There's another piece, the one that looked like aluminum foil, which resembled the inside of a Philip Morris cigarette box, which was a light piece of aluminum. Can you describe that piece? Or what you did with that piece?"

Reme: Well, when we first started down the road to that crash site, this piece was under a rock and I saw it glitter and it was moving up and down, and so I pulled it out from under the rock and kind of rolled it up and folded it up and you would fold it and it would go back into the same position as it was. Today they call it "memory metal", but at that time we didn't know what it was.

A cautious word about technical history is important here because, as in the case of the transistor, many UFO books have incorrectly described how certain inventions came to be. While the transistor effect had been known since the 1930s (and an old German patent proves it), and therefore was not derived from the Roswell crash, the same is true for memory metals.

As early as 1932, a scientist named Arne Olander had discovered the "pseudo-elasticity" of an alloy of gold and cadmium. Six years later, in 1938, Greninger and Mooradian had demonstrated the effect in a copper-zinc alloy, although it wouldn't be before 1949 and 1951 that the "memory effect" became widely reported. The modern nickel-titanium alloys like Nitinol were discovered by accident and developed in 1962-63 by the Navy.

This shows that, while the physical concept had been known for years, no practical application of the type described by our witnesses would have been realistic in 1945, let alone in a craft with the structure and capabilities of the one that crashed on the Padilla land. It was several years too early in practice, even if the concept was already there in science.

Reme Baca, therefore, is justified in saying that "at the time, we didn't know what it was," and he goes on, still talking to Paola:

Reme: I took that and put it in my pocket and took it home, and showed it to Jose and played around with it for a while. I had a can in which I had a few Indian head coins that I was saving. And so I put it in that and then I took that can of Prince Albert and put it in the well. And it was there for quite a while until one day my dad had come home on vacation. We were lucky if he came home once a month, because he worked in Albuquerque at the Veterans Hospital.

When he came home, he was working on the windmill and he said, "I can't fix this thing. Maybe you and Jose can take it and get it welded." So we took it apart. I talked to Jose, and so we went in his pickup to Socorro and took it to a blacksmith, and asked if he could weld it. The threads were worn. And so he looked at it and he says, "You can't do that, because it's made out of brass and I have no way of welding brass".

So that didn't turn out very well. We came back to the well to see what we could do with it. We'd tighten it and then it would slip. So in an act of desperation I said, "Jose, reach back there behind that piece of wood and there should be a can of Prince Albert there." He reached back there and found the can. There was that little piece of aluminum type foil that we had found at the crash site. Maybe this would work.

We took that aluminum-type foil and we wrapped it around the threads and then we tied the ends tight and it didn't slip, so then I took one of the Stillson wrenches (*a type of adjustable pipe wrench*), Jose got the other one, and we tightened it up. We went and turned the windmill on, and it started pumping some water. It worked!

Paola: It worked?
Reme: It created a vacuum, and pumped water.
Paola: Very, very ingenuous thing to do with a piece of extraterrestrial material... (*Laughs*)
Reme: We didn't know. No way of knowing! And so as time went on, Jose left there, and then I left there, and went to Washington State

Chapter Four - The Secrets are Kept

and went to school and then I eventually got married, and we were raising a family. Sometimes both my mom and dad would visit us. This time, my dad came alone to visit and he stayed at our house overnight. Ginny fixed lamb chops for him. That was his favorite meal. While we were sitting at the table having dinner, my dad turns over to me and says, "Reme, I don't know if you remember or not, but you know that cylinder that you fixed on that windmill? It's still working. Don't know what you did to it!" I responded that I was glad it worked out. I couldn't bring myself to tell him the real story, because he wasn't supposed to know...

Paola: ...where that came from?...

Reme: ...that this was some form of advanced alien technology at work. He would never have believed it. And he wouldn't believe that we have phones today, cell phones that you can carry in your pocket. And so, that would have been a lot for him, so I just thanked him,

"Great, I was glad to do it."

^ ^ ^

Much of my research about UFOs has involved field research on three continents. In that process, I was aware of some 40 or 50 reports or claims of crashes and recovery over the years, but few were documented, so I did not become seriously committed to the study of those materials until enough pieces of researchable hardware from the most solid cases had accumulated in my own files, sent by field investigators or by the witnesses themselves, with little or no publicity.

In that process, I found a precious colleague in Paola Harris, who had spent as much time as I did studying the official files and—more importantly—meeting the witnesses themselves. But while much of my work covered France and the United States, she had researched the most significant cases from her office in Italy, and later conducted extensive field work in the course of many trips to Latin America. **(15)**

As a result, we discovered that between us we had achieved a fair global assessment of the nature of the phenomenon that reflected the massive, undocumented reality of thousands of unexplained observations "on the ground" on every continent. It transcended the local North American obsession with military cases, pointless efforts to get money from Washington and never-ending media quarrels about unproven conspiracy theories.

We could only laugh or cry at the wild rumors on every cable channel and late-night radio about supposed American bases on Mars and the oddly-designed "Serpo" hoax that drove naïve researchers into a maze of blind alleys from which they haven't yet emerged. **(16)**

Among Paola's most notable investigations are her long-term association with research in Canada through the good offices of the Honorable Paul Hellyer, the former Minister of National Defense (Fig. 10) and through her work reviewing and documenting the experiences of Colonel Corso, undoubtedly the most knowledgeable US Army officer covering the field in an official capacity in the period that interests us in the present book, namely the end of World War Two and the beginnings of the Cold War. The Italian version of Corso's book, which Paola guided through publication, was fortunately free from the constraints placed on him in the United States by his agent and his publisher, a form of censorship that had infuriated the old soldier.

In both of these areas, Paola and I had long-standing personal connections based on trust, and we were familiar with the same dossiers. I had spent time with Colonel Corso both on my own and with colleagues in Las Vegas, and I also kept a French-language database of cases in Canada, where I had made multiple professional trips in connection with technology investments.

Science has little to do with politics, advertising or religion. It doesn't start from belief and it isn't done by lobbying Congress or by gathering and amplifying rumors from other people: it begins with individual research on hard data and reproducible experiments. So I followed what other scientists and private researchers were doing in the area of recovered materials, I listened to their presentations and I tried to make sense of what they found, helping whenever I could.

Some stories left me skeptical; others, particularly about Roswell, left me confused and a bit disgusted; and still others fascinated me.

They still do, but things have moved on: Now we have real data, and the means and the tools to bring research to a much higher level.

∧ ∧ ∧

The reader may be distracted by recent claims, in the press and on TV, about flashy ideas presented as "unique breakthroughs" or imminent disclosures about Aliens. Such is intellectual life in America in this new

Chapter Four - The Secrets are Kept

Fig. 10: Paola Harris, reviewing research data with the Hon. Paul Hellyer, Canada's former Minister of National Defence, who knew the work of Wilbert Smith Toronto, Feb. 24, 2006.

Fig. 11. Colonel Corso and Paola Harris on a panel in Pescara on the Adriatic coast (1998) She sponsored the publication of his book in Europe, where he was able to reveal key data that had been discouraged from publication in the US.

century: colorful, superficial, aggressive, cleverly marketed, boastful, occasionally brilliant...and too often misleading.

Much of it is driven by motives that have more to do with political and publicity convenience, or just plain, opportunistic greed than with genuine research. It corresponds to a desire to influence the belief systems of younger generations, rather than a serious effort to preserve, and build upon, the accuracy of facts. And its reasoning is wholly based on shortcuts.

Dangling the promise of "UFO Disclosure" is as good a marketing strategy as any, since progress cannot be measured by any realistic standards, and there is no defined end point to the research.

It is also, largely, an empty promise because every segment, group and cult among the colorful crowds at UFO conventions defines "Disclosure" differently. Public reactions to exposure of the reality of the phenomenon, combined with ignorance of its nature could have unpredictable and confusing effects, to the dismay of officials.

Understandably, the academic community has seen through the charade and generally declined to get involved. But without the depth of knowledge and the open debate tradition of the modern academic world, very little real progress will be accomplished in this field. Because the old theories won't work.

That confusion in the modern media was one of the factors that discouraged Reme and Jose from going public with their data.

The most obvious negative effect of the circus atmosphere created by clamoring voices on cable TV and web podcasts (with a few laudable exceptions) is the quiet withdrawal of those witnesses and of researchers who possess real, often scary or disturbing evidence. The fact that the two witnesses of the San Antonio crash remained silent during their whole active lives is a prime example of this, but I have others of equal relevance in my 500-odd files of over 50 years. This should tell us something about the chances of making real progress by parading witnesses on TV or re-casting their real experiences as flashy studio recreations, ignoring the genuine enigma and any complexity if it doesn't fit nicely between two commercials.

In the real research world, the serious study of unexplained materials from UFOs has a long, quiet and distinguished history. At Stanford University, Professor Sturrock has conducted and published analyses of the famous Ubatuba explosion of an unknown device in Brazil **(17)** since the 1980s. Other researchers within groups like CUFOS and MUFON **(18)** also conducted serious field research and chemical analyses long before the turn of the 21st century. I had occasionally contributed to that work and I had been invited to assist in official studies of similar cases with French, Brazilian and Russian investigators, but my main research interests remained elsewhere. When I published the first peer-reviewed assessment of UFO cases with hard physical data in 1998 **(19)**, it was done from the point of view of a global search for information patterns and energy parameters, rather than as a fresh investigation of the primary data in the field – much of which, frankly, left me skeptical at the time.

The opportunity to re-investigate the 1945 San Antonio case, which appeared to me as the main "pivot" in the field, has made me re-enter the problem from a different direction. When Ron Brinkley and I began our field work on "The Plains" thanks to the tireless efforts of Chuck and Nancy Wade, and when I became more aware of the local history of recovered materials, I began looking for guidance from earlier researchers. Mr. Steve Murillo, of the Paranormal Research Society in Los Angeles, encouraged me in that direction. It was Mr. Murillo who introduced me to Paola Harris in July 2018. I then became aware of the deep investigative work she had done over the previous nine years, when she sponsored some of the field studies of independent researchers and actively networked with high-level scientists and government officials in the US and other countries.

An associate of Professor Hynek for many years, Paola had gathered extensive files about cases in her native Italy, and her contacts in Canada gave her access to the historical records going back to the scientific studies conducted by Wilbert Smith, decades ago, about which I had my own information sources.

Similarly, Ms. Harris had done extensive work with Colonel Corso, the man the Army had entrusted with the testing of recovered hardware from crashed UFOs in the 1950s and 1960s.

Thanks to the opportunity to work with Robert Bigelow's National Institute for Discovery Science, I had spent two fascinating days with Colonel Corso, some of it in private, and he had told me his concerns about some unpublished experiences he'd also mentioned to Paola. This sharing of the inside story of the early crashes built a bond of trust between us.

The story of the Padilla Ranch was *NOT* just one more ufological rumor, amplified through some cable series or reformulated in slick, trendy tones by the modern media. This was real, hard-core, historically significant data, and it had never been brought to the light of day on a consistent, scientific basis. It was the stuff TV channels didn't dare to touch because it might antagonize some sponsors, and the inconvenient reality that academic pundits were wary to take into account, because there was no financial support to pursue such "frontier" work.

While we started sharing data over the summer of 2018, I didn't meet Ms. Harris until the San Antonio case gave us an opportunity to work together. I was able to travel to Albuquerque in October of that year, and I then started learning from the storehouse of information she had quietly amassed in previous years.

Much of it centered on the nearly-forgotten Padilla case, which we were eager to bring to a new level of analysis.

As we reviewed the transcripts, the maps, the early video interviews and Paola's own extensive collection of photographs and tape recordings, it became obvious that the case had matured over the years, in spite of the long separation between the two witnesses--or perhaps because of it.

There had been no publicity, no big story, no interest from TV reporters and only casual attention from a few ufologists. Paola had the only consistent, historical data and was eager to upgrade the research.

Today we can marvel at the fact that the significance of the event was missed for so long, although it took place close to the first nuclear explosion, in time and geography. The strategic questions were obvious: This wasn't just the ordinary UFO case, a stranded motorist or a passing tourist seeing a strange light. Instead, could it be a response to the bomb? The start of a dialogue? A warning?

Before we could entertain such speculation, there was a lot of work to be done. Some of the statements by Reme Baca, for instance, were not clear to a scientist. There were contradictions in the testimonies, yet to be clarified. And the site itself was changing with the years, the floods, and occasional interventions by the Bureau of Land Management, from which the land was leased.

Some of those official actions did not make much sense, either.

After removing the disabled craft, for example, government teams came back twice: The first time in 1953, apparently to look for any evidence that the military had missed, or any piece of the object that might have fallen from the trailer as they drove away from the property over the dirt roads; and the second time in 2017 when civilian employees planted intensely poisonous vegetation precisely over the area where the craft had come to rest after careening down the slope, spilling more debris.

Chapter Four - The Secrets are Kept

Those official actions seemed bizarre to us. Were we too suspicious of the official desire to maintain secrecy? Or were we still missing some key signs of the significance of the crash, its relationship to the dawn of the Atomic Age? Or was there some piece of advanced technology they were missing, along the lines of the classified speculation that would surface in the following years about quantum effects, dark energy and "warp drives"?

When we discussed this aspect, it became obvious to us that everybody might be using obsolete hypotheses. While much has been made of research into complex composition of UFO residue and isotopic ratios, the naked truth is that *very advanced technology is not necessarily esoteric*: Automobile propulsion beyond today's much-advertised electric cars running on environmentally-disastrous batteries may involve nothing more than hydrogen, the simplest and most abundant element in the universe, and produce exhaust in the form of pure water. So why are we posing as a principle that Alien vehicles must use horribly complex materials?

The only way to bring new light to that old problem was to continue our research, with no preconceived assumptions about extraterrestrials or contemporary theories of physics.

The time had come to put the documents aside and to spend time at the actual site with Mr. Padilla, a man who had seen it all happen on his family's ranch, seventy-two years before.

∧ ∧ ∧

PART TWO

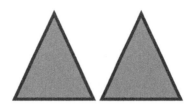

STALLION GATE

*No man-made phenomenon of such tremendous
power had ever occurred (...)
The whole country was lighted by a searing light
many times that of the midday sun.*

Brigadier General Thomas Farrell, 1945

CHAPTER FIVE

FIRST SITE VISIT, OCTOBER 2018

On October 12, 2018, a Friday, I flew to Albuquerque and met Paola Harris at the airport. As mentioned earlier, it was Steve Murillo, of the Paranormal Research Society in Los Angeles, who had put me in touch with Paola Harris and her team a few months before.

Although that was a first visit for me, Paola had been there four or five times before, and was well familiar with the confusing landscape and the best access roads.

We drove straight down to San Antonio. I recalled that Reme's wife Virginia ("Ginnie") had astutely captured her impressions of the area when she visited it with her husband:

"It was surprising the number of adobe houses one encounters in San Antonio in keeping with the truly deep heritage of New Mexico. San Antonio has its share of Frame, Stucco, Brick and other types of houses. The adobe bricks were made right there in town at very little cost. San Antonio is a very small town in central New Mexico, steeped in traditional, conservative values, beliefs and attitudes of the community. It seemed to me, that Jose and Reme's discovery of an alien craft, by its very nature, would have challenged those conventional beliefs and attitudes. It may not have been acceptable to them either, for it would demand to be confronted, and once it was, it would be impossible to dismiss."

Paola introduced me to Jose Padilla at the Owl Bar. Nothing had changed there since my first visit with Ron Brinkley, exactly one year before.

Mr. Padilla is a few years older than me, an outdoor man with longish grey wavy hair and an active, friendly approach to people and events. He's also a retired State Trooper of long experience, and a quick judge of people. I must have passed the first test, because he inquired about my background and the state of the research, so I told him about the mounting interest towards the UFO subject among some of my scientific colleagues, with a renewed focus on hard data, recovered material samples and other physical evidence: we now had access to extremely fine methods for testing materials, I said, with improved instruments that had been either too unreliable or too expensive to be used before. Some of them had only been invented and perfected to actual practice in the last couple of years.

The waitress brought our meals, with a "chili dog" for me, an old Southwest tradition I'd missed since my days in Texas, to the visible amusement of Paola who wasn't used to Frenchmen eating sausages and black beans swimming in spicy Mexican sauce.

Jose spoke to us about his experience as a child during the intense days of World War Two. He confirmed that the scientists of the Manhattan project used to come to the Owl Bar for food but also to buy common staples: It was the only basic market for many miles around. Security in the area was intense: When they saw the serious men, always carrying leather briefcases full of papers, other customers thought they were business types, perhaps insurance executives. The Army counter-intelligence agents may have encouraged that idea, and it would be many years before the locals realized they had eaten their meals in the same room as Robert Oppenheimer, Hans Bethe, Enrico Fermi, Phillip Morrison or Edward Teller, the princes of international physics. Von Neumann himself had been there. And yes, that dilapidated structure across the street was the ruined compound where some of the scientists stayed at the time of the Trinity atomic tests.

Mr. Padilla agreed to take us to the crash site the next day. We made arrangements to meet him in town and to drive together to the ranch where it all happened.

∧ ∧ ∧

Chapter Five - First Site Visit, October 2018

Fig. 12: Paola Harris at the crash site with Mr. Padilla on the Ranch in 2016.

Fig. 13: The same site in 2018, with newly-planted poisonous vegetation.

A typical New Mexico storm was threatening, as it was on the day of the crash, when we got there after a short drive south of Socorro on highway 25, the historic Pan-American Highway, to the small cluster of farms and ranch houses known as San Antonito. We reached a wide gate, turned into a dirt road and Paola's rented SUV danced up and down on the rugged terrain for what seemed like a long time before we reached a spot where we could park safely on dry land in case of sudden, torrential rain. Then we left the car and looked at the landscape.

What I saw was a gentle valley between two rows of hills, leading to a hollow, oval spot where strange, yellow weeds were growing, about three feet high. Knowing about the notorious New Mexico desert floods, however, I could imagine days when that scenery would be far from gentle. As for the tough yellow plants, I couldn't recognize them.

In the wide cleavage between the hills, the bluish horizon was clear under the grey, tormented sky. The three summits of the Trinity Mountains were clearly visible, like inverted V's in a short line: from the northeast to the southwest: Oscura, Little Burro, and Mockingbird.

Between us and Trinity, Ground Zero was lying lower in the desert, invisible to us.

We began taking the measurements we needed, understanding the lay of the land, planning our next steps. Our initial idea of surveying the actual crash site, and digging there, was obviously unrealistic, given the presence of the dense, poisonous vegetation.

We were standing about 40 miles away from the three peaks and only 25 miles or so from Ground Zero. Jose Padilla pointed out the curving trajectory the object must have followed to hit the Marconi tower some 2,000 feet away and come crashing to the area where the yellow weeds were growing.

How nice of the Government, I thought, to mark the exact spot for visiting ufologists! But they could have marked the spot with more elegant flowers. (Note)

Mr. Padilla told us there had been small flowers there after the crash. The spot had remained rather bare, but in time one could see a wide circle

---

Note: There were small pieces of the tower left on the ground when Paola Harris first went to the site, before Major Frank Kimbler removed them. Later Ms. Harris found an electrical connector that came from the wrecked tower.

of small flowers. This was the first ground-level observation that we could relate to other cases I had studied, where such large fungi appeared to have been stimulated by an external agent. One of the most celebrated cases I had examined was that of Delphos in Kansas, where such a plant, essentially a type of mushroom, growing out from an underground spot, with its fruiting bodies at equal distance from the center, had become prominent after a UFO "landing."

Botanists who did the determination of the exact plant for us told us its rapid growth in the Delphos case could have been stimulated by electromagnetic radiation. After the incident the plant actually fluoresced, a fact the scientists told us was quite rare.

In medieval times, the phenomenon of these large radiating mushrooms gave rise to stories, legends and tales about "fairy rings."

^ ^ ^

"There was a rainstorm the day it came down," Jose told us. "We heard something like a sonic boom, and we saw some smoke. We got on our horses. The thing had hit that very high tower, then it spun around, descended to this area and crashed, leaving a long gash in the earth, as long as a football field. The greasewood caught fire. The closest we came that day was about 300 feet. There was a hole; *we saw small men running back and forth as in a panic*. We must have been there two hours, and then we went home and we told my father. He called the State Police."

"We went back two days later but most of the scattered stuff had been hauled away, except for the craft itself. The outside shape was an oval, some 30 feet long, 15 feet high, with a little dome towards the back. The color was greyish rust, not a real color. We saw the whole recovery operation."

"On the last day the object was under a blue tarp, ready to be hauled away, so we waited for the guards to leave and I sneaked in. Inside, it was like yellow brass, smooth but not shiny. I saw a panel there, about 30 by 24 inches. It had a copper ring, it looked like an electrical device. A pin held the thing together, with a dial that could swing around."

He pointed to the spot with the tough yellow grass (Fig. 13): "This is the area where it crashed, after that wide arc following the hit against the tower. It's an ancient traditional place for the Indians, by the way; an Apache sacred spot."

We walked closer to the crash site. The plants reached higher than my knees. Paola, who'd had them analyzed, said the common name was

"Cockelspur." She warned that touching them triggered an intense allergic reaction with immediate, painful effects; a sure deterrent to anyone bent on searching the area for any traces of the crash. Nobody seemed to know who planted them.

I could only concur with the observation that the selection was deliberate: If the Federal Bureau of Land Management had wanted to prevent erosion or mud slides, they could have used many other types of plants.

∧ ∧ ∧

"As soon as they realized what we'd found, The Army Air Force came to our home," Jose said in response to our further questions. "They asked my father for permission to cut the fence and build a road, "to take out their weather balloon."

"I was only nine years old at the time," Jose went on drily, "but I already knew, that thing was no weather balloon!"

"We were told not to go back, but the crash was on our land, so we went back anyway. We knew how to hide from the soldiers. They did cut a special road to bring a large flatbed truck; then, they built a big A-frame to lift the thing." **(20)**

Jose and Reme quickly noted that the military didn't always keep a secure watch around the object: the young soldiers took every opportunity to drive out to the Owl Café for meals and beer. Then, the two smart kids would get out of hiding, as Mr. Padilla told me, to sneak another look around the unknown craft.

Most loose objects had been removed, and the distraught small beings they had first seen staggering around had escaped, or they had been caught and taken away, but there was an intact panel attached to the inside curved wall, with a metal device that swivelled; so Jose confirmed to us that he'd daringly wrenched it out:

"I told Remigio to go and get a cheater bar from the Army truck," Jose told me, smiling at the memory, "and I took that thing away!"

Back in town, unbeknownst to his father, they hid it under the floorboard of a cabin where farm hands occasionally slept.

"The military came to ask my father if he was aware of anyone stealing something from the object, and he said no. But a sheepherder who slept in the cabin was scared one night by 'hombrecitos' who said they were looking for the 'tesoro'... He related the story that morning as he came to the main house, still shaking from the encounter."

The kids kept their mouth shut, their father didn't believe his employee's story, and the metal bracket remained a secret. Jose recently gave the piece to Paola, who cut off a small section for me to have analyzed.

It had already been tested in Mexico (on October 24, 2017) at the Centro Educacional Analitico where chemical engineer Bernabe Hernandez Santos found a preponderance of Aluminum (757 mg per gram) followed by an 'unreadable' element for 34.6 mg, then zinc (27.0), iron (6.5) and another unknown for 158.5 mg. All of which didn't quite add up.

"I don't believe in unreadable elements," I told Paola. "There must be better results somewhere and anyway, we'll redo the measurements."

^ ^ ^

Bad sleep, and worry: Why and how did Ron disappear? I wondered, as I tossed around in my bed in Socorro. He should have been here, enjoying this investigation he'd inspired, adding to it with his great patience and ability to note details in context. Paola had agreed to stop again in Albuquerque on the way back to the airport, and to visit the building where he lived.

"Perhaps he's moved away, or suffered some accident?" I thought. I didn't want to believe he might be upset with me about something, he would already have told me so.

The next day, Paola did drive us to Coal Avenue where residents confirmed Ron had rented a room in a rectangular, single-story building arrayed around a courtyard. We went around the complex again. A woman came out of one of the apartments, so we asked her about Ron and she flatly said he was dead: Two months before, before dawn, riding his bike to his job at an airport shop, he had been hit by a speeding truck. All his belongings had been sent to his brother. The apartment had already been rented to new people.

I went to the shop where he worked as a storeroom employee. They said they weren't in contact with his brother.

I had feared this. New Mexico is a tough place and it can get ugly, even under the bold display of its cosmic beauty. I had thought of a car accident on the mountain roads he drove on his second job, distributing newsletters on the reservations, talking to local folks, listening to their stories. I didn't know he rode his bike to the airport.

I told Paola about Ron, our friendship of many years; he'd known my wife and my kids, he stayed with us when I built a small observatory in Mendocino County; we went on many expeditions together, and I admired how sharp and original his mind was.

As I waited for the flight I bought turquoise cufflinks from his shop to preserve a sign at least, of everything he meant to me. When I got home I found this letter from his brother Ken and Ken's wife Melissa, responding to my earlier enquiry:

"Your concern was apparent and true. Ron was involved in an accident on August 6th. He was on his bicycle going to work and was struck from behind by a motorist. Despite wearing a helmet, he lost his life due to extensive brain damage. It was a hit and run, but they found the driver and he will be on trial for what he did to Ron, and the callousness of leaving him on the road that way."

∧ ∧ ∧

CHAPTER SIX

GROUND ZERO

Our next visit to the area was supposed to happen on April 6, 2019. The day began with an interesting coincidence that precipitated a change of plans.

We had agreed to meet for breakfast at the motel coffee shop and to drive straight to the crash site, so I got up early and waited for Paola and her able assistant, Bill Crowley, an investigator from Los Angeles who had field research experience. They were late and I was hungry, so I got my coffee and I looked around for something to read. Someone had left the day's paper on a nearby table. My eyes fell on the front page, with local news about a political demonstration: A group of environmental advocates would stand by the side of the road leading to Ground Zero to protest continuing work on nuclear weapons.

And why were they going to mount such a protest in the middle of an empty desert? Because this was one of only two times during the year when the Atomic Site was open to the public, so cars were going to line up for miles to take a turn at that particular intersection.

Full article on page three.

When my friends showed up, I told them there was a change of plans, and I took the wheel on an unplanned drive to Trinity.

∧ ∧ ∧

Once you leave the main highway, veering off towards the old Army base, the landscape unexpectedly embraces you in silence and majesty. You're driving along the immense seabed of a prehistoric ocean, with no other boundaries than the far-off mountains, hours away, dark and fuzzy in the early heat against the uniform blue sky. You see the wide expanse of White Sands to the south and only the vague outline of the Sevilleta Range (a National refuge) far away to the north. Thirty miles ahead is the Chupadero Mesa, lost in the haze.

"The Plains" capture you with their secret beauty and vastness, even if at first their uniform appearance has seemed uninteresting, unadorned by any trees or human points of reference. A sense of eternity eventually takes hold, squeezing your heart and lungs at that altitude, suggesting mystery—or unseen danger in the thin atmosphere.

To relieve the vague anxiety weighing on us, I turned on the radio, rewarded by a Spanish-speaking music program. Pedro from Las Cruces dedicated a song to his new girlfriend, so we were treated to an accordion and trumpet duo that restructured the desert with hopeful melodies of love and the sorrow of separation: *Cuando podria verte?* When will I be able to see you?

Somehow, it all fits together, as you relax and keep watch on the odometer.

Twelve miles later, however, you still haven't seen any sign of man or any new feature, and you begin to realize that something isn't quite right with the sparse vegetation and the powdery dirt. The dwarfish plants themselves look funny: *madrones and manzanitas should be taller; there should be real bushes; there should be a lot more grass.*

Now people keep arriving in mobile homes, cars and motorcycles. Families with kids, tourists, a little dazed like us, disoriented in the flat uniform landscape with its strange vegetation, its abnormally short mesquite bushes.

The mariachi keep playing. The announcer comes on, cheerfully telling Maria from Santa Fe that Francisco, on a Navy destroyer somewhere, dedicates a song to her on her birthday. Then it dawns on you there's a reason for the degenerate plants and the stunted trees: The landscape was irradiated, seventy-five years ago, by the deadly light from Trinity. Hadn't Dr. Oppenheimer, watching the test, quoted the Bhagavad Gita: "I am the Destroyer of Worlds…"?

More realistic, less poetic but no less impressed, Brigadier General Thomas Farrell had written: "The effects could well be called unprecedented, beautiful, stupendous and terrifying."

The immensity of the devastated plain we're driving through is silent testimony to the General's attempt at an account of the awesome spectacle: His statement still resonates with me when I recall the empty sky and decaying vegetation.

∧ ∧ ∧

It's a relief to reach Stallion Gate where an unarmed guard in fatigues, a cheerful teenager – the first human we've seen in what feels like a long time – waves us forward to join the flow of cars, pickups, and even the odd official limo with tainted windows, stuck in traffic ahead of us.

The massively black, silent, air-conditioned contraption, with its antennas, its driver in conspicuous black suit and tie, and the hidden silhouettes of its occupants seemed as bizarre, in this landscape, as a spacecraft from Dimension X.

The young soldiers at the fence check our IDs and wave us forward into the desert beyond, with more miles to go before the parking lot. To the right (North) is another road leading to the McDonald ranch where the scientists had assembled the plutonium bomb. To the left, another wide fence perimeter and beyond that, a black obelisk, awkward and stocky, marking the spot of the first explosion.

Like most military sites this one is simple, basic, the various parts documented in plain English on a few posters – and it seems all the more terrible in the weight of its historical significance: Here, in July 1945, humanity was thrust into a new, terrible world from which we never emerged.

American crowds, so diverse in their composition, motivations, dialects and behaviors, often seem united by a restrained fervor that links the Hispanic families, the couples in Sunday best and the joking students, the rough and ready motorcycle toughs, the tourists and their cameras with

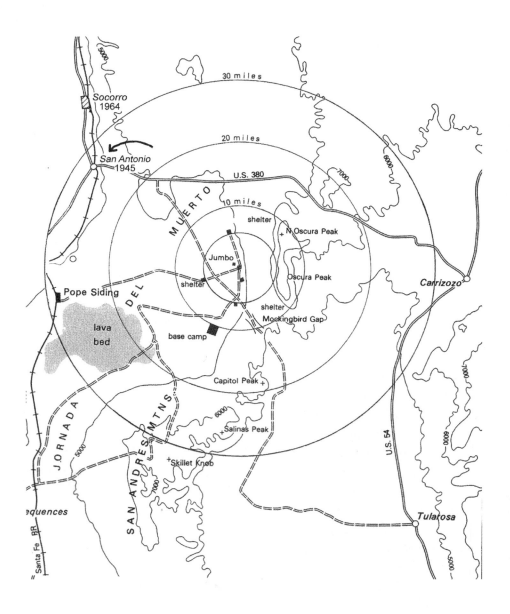

Fig. 14: Detailed map of the Trinity Site, with probable trajectory of the object, as reconstructed by Mr. Padilla.

Chapter Six - Ground Zero

fancy lenses, among black families, brown neighbors and Asian visitors, and of course the three of us, Paola and me and her assistant, on a research assignment from a mysterious force—or an unknown cosmic joker. But the speculations evaporate when the kids in camo direct us to a vast parking lot a few more miles away. They obviously enjoy the one-day relief from routine desert duty.

An entire city is assembling in the dust.

∧ ∧ ∧

Mid-morning: hundreds more arrive; vehicles of every make, shape and age maneuver in every direction. We look for markers, and I take a few orientation pictures, so we'll be able to find our own car again in the afternoon. Then it's time to follow the lines of visitors wandering off towards the "Marker," that ugly pyramid, really just a pile of black stones, the Army's idea of great landscape art, where the Tower once stood.

No vegetation at all here.

Fig. 15: The enclosure of "Fat Man," the Nagasaki plutonium bomb.

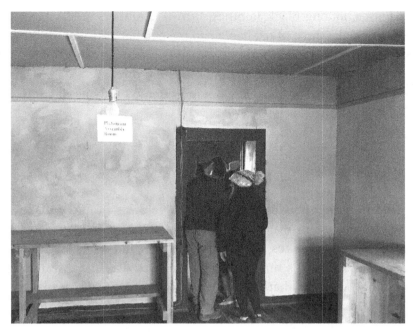

Fig. 16: Visitors to the Plutonium Assembly "master bedroom" at the old Schmidt-McDonald Ranch.

At 5:29:45 am MWT ("Mountain War Time") on July 16, 1945, the explosion at 15,000 degrees Fahrenheit, a level of heat equivalent to that found at the surface of the Sun, destroyed everything, melted the steel tower, made a flat depression in the surrounding ground and even created a new mineral, "trinitite:" a shiny, greenish substance resembling molten glass.

It would be years before the details of the actual process of the formation of trinitite were finally understood, because the heat of the blast melted the desert sand in a peculiar way, never seen before and actually impossible to reproduce with a natural process.

In a report published in the Fall 2005 issue of the *Nuclear Weapons Journal*, scientists Robert Hermes and William Strickfaden finally explained what had happened: Instead of being baked under the blast, as one would have expected, much of the sand was first scooped up into the fireball, all the way up in the atmosphere, where it behaved as water does in an ordinary cloud: Tiny droplets aggregated into larger droplets and they precipitated back to Earth as a rain of molten glass. The fireball then fused them into a smooth surface.

Chapter Six - Ground Zero

I silently recalled how Enrico Fermi, that morning, had done the first evaluation of the energy achieved by the blast. Then he'd climbed into a special Sherman tank covered with radiation-absorbing lead (Fig. 5) and explored the half-mile depression the bomb had dug up in the desert where we now stood in silence while visitors scattered around us over the wide area enclosed by barbed wire. Like ourselves, they wandered in quiet awe from the model of *Fat Man* (actually, the casing for the 10,800 pound, complicated plutonium device dropped on Nagasaki, see Fig. 15) to the amazing steel cylinder of *Jumbo*.

It had been just an unplanned side trip, taken on a whim, triggered by the random finding of a newspaper someone had left behind at the motel coffee shop, but our visit to Ground Zero isn't an experience I will forget as long as I live.

It also provided us, in a most timely way, with the ideal template to better understand what Jose Padilla and Remigio Baca had seen on a peaceful ranch, one month after the Gadget turned Ground Zero into a flat lake of pale green glass.

∧ ∧ ∧

CHAPTER SEVEN

SECOND SITE VISIT, APRIL 2019

When we met the next day at the hotel, strangeness hit as soon as Paola and Bill Crowley greeted me. We had just sat down for a snack at the restaurant. Bill had begun telling us where we could rent the equipment we were going to need (magnetic detectors and a standard radioactivity sensor, primarily), when a man walked over to our table and shook hands with Paola. He was George Waldie, a retired friend of a geologist called Richard Sigismond, himself a well-known researcher who had worked with Dr. Hynek.

Sigismond was a friend I had initially met in Boulder. He had been dead a few years. We shook the suggestion that these presences were coming back from the Other Side to be with us... What brought George Waldie to New Mexico, and why did he expect to find us in that particular restaurant?

The episode was never clarified, but it was the second bizarre incident during the trip, and it cast a disconcerting shadow over our little expedition as we headed for the parking garage.

Buying a shovel and a hoe, renting two detectors and piling up water bottles in the back of the car took a couple of hours. In the process, I got to appreciate Bill's energy and insight. His trajectory in life had common points with that of Ron: He too had spent time fishing in Alaska, loved to be alone in nature, volunteered as a watcher in forest towers to alert helpers in case of fires or lightning strikes; he'd spent time in Japan and understood cultural subtleties. He was also a member of a research group in Los Angeles with investigative experience.

We visited several shops in Albuquerque to gather the equipment we needed, also some food and drinks, before heading south again, to the Padilla Ranch.

^ ^ ^

There had been a recent flood in the area. It had damaged the roads. Now we were driving off to meet Jose and to visit the property again. Perhaps the flood would have brought more pieces of foil to the surface, we said hopefully; but it might have buried them even deeper; or carried them downstream for miles, never to be found.

I asked again about the tough yellow weeds, where did they come from? And who planted them, if they were not native to the area? They

Fig. 17: Discussing the case with Mr. Jose Padilla in San Antonito at the 5000 foot level.

Chapter Seven - Second Site Visit, April 2019

Fig. 18: Searching the bed of the arroyo for residual samples.
(video record by Paola Harris)

weren't there on Paola's earlier visits, when she had taken researchers Lori Wagner and James C. Fox to the site.

Other researchers had asked the same question, years before us. Paola told us that in 2012 she'd introduced Mr. Padilla to independent researchers who also went to the site with her and became intrigued by the same problem. The plants formed a perfect oval where the craft had careened to a stop in 1945. That oval was about 30 feet in size: "For seventy years, nothing would grow in the center and there were tiny little flowers around the edges," according to a report Paola made available to me. "We decided to take soil samples in the center last year and we had those soil samples, which all dried out. Nothing significant was in those samples then."

It was in July 2015 that the location was found transformed, with the spectacular thick growth of the tall, rugged plants the reader can see on my own photograph (Figure 13). Paola and the group then assembled a team that included David Garcia, who was involved in collecting samples, and Tom Hamlin who had been the first researcher on the case in 2012. (Note)

This was a respected team of knowledgeable researchers and we needed to learn from their past experience. Some of their findings were detailed and quite useful.

Note: See Paola Harris: *Dawn of the Atomic Age*: Mystery at San Antonio, New Mexico crash site. *MUFON Journal* no.578, June 2016.

"The initial oval landing site (...) does collect water but dries up immediately," they reported.

On previous times, Paola had found the spot barren: there was nothing growing in the middle but the new report said: "This year, for some reason, there were these unusual-looking plants in the middle, plants that are not seen anywhere else, neither on the Bureau of Land management fields, on the land that is leased to Jose Padilla and his family, nor on their 80,000 acres. Nowhere near the Owl Bar and Café or anywhere nearby have we seen those plants. The plants are poisonous and very, very irritating. My assistant Tom Hamlin picked up the plants with his hands and had an allergic reaction immediately. His face became red and irritated. Areas on both of his hands and arms developed hive-like irritations, and we very quickly saw that we can't touch these plants. The irony is that the plants are only within that circle. They are only within that 30-foot diameter circle where the craft landed."

The team had cut some plants out, wearing Hazmat suits covering their face and body with glasses, tape, gloves and boots. They also took water samples. Unfortunately, that first analysis was delayed and the results were incomplete, with only some speculation that the plants might be Jimson Weed, which turned out to be wrong.

Those are the things you go through when you conduct a serious investigation: What had seemed obvious vanishes. Other obstacles come up, unexpected, nagging, and sometimes dangerous. They waste your time.

I must take the reader through those details because those are the things you won't see on TV, with slick scripted dialogues among intrepid researchers who always dig at the right place, or exhibit impeccable graphs from their field computers.

The reality of field investigation, even when you're guided by people who were there at the time and know the territory, doesn't conform to any neat pattern. You have to follow your intuition and invent new procedures; then you have to check everything and do it again.

Paola Harris had returned to the site with Australian researcher James Rigney to gather fresh samples, and this time the analysis discredited the Jimson Weed idea. Instead, a DNA study conducted by botanist Dr. R. Jason B. Reynolds at Colorado State University led to positive identification. So did a parallel study by the BLM Socorro field office.

"The Park Rangers have identified these plants as *Night Shade* and *Xanthium strumarium (Cocklebur),* just from the photos we took. We are still waiting for formal analysis," the report said. Evidently the plants had been purposely put at the site.

Both types of plants were poisonous, as it turned out: the green one (the cocklebur) and the silvery one, which was the nightshade, cautioned the report from the Park Rangers, still working from the color photographs, far away, in the safety of their offices.

"What is underneath the soil within that 30 foot diameter?" someone asked. The seven young soldiers who were assigned to cleaning the site in 1945 had thrown many of the small pieces into a trench, as Jose had told me during my first visit. But when I was told, "*All it would take* is a bulldozer digging 20 feet down beneath the ground..." I had to laugh.

To me, the idea of digging was unrealistic for several reasons, including the plain requirement to maintain slope stability in the area, and also because that would take a Federal permit, and formal clearance from BLM. Another blind alley. There would be others.

We gave up on the idea of recovering artifacts at the site.

^ ^ ^

After all that discussion of the extent and evil purpose of the plants, we had a surprise as soon as we drove up to the site: The earthen dam had been strengthened and built up against the latest rains, and half of the oval area where the plants had stood on my first visit was now flattened to the same level as the ground upstream.

Over that entire zone, the evil plants had been removed. The result of the levelling, however, was that any remaining pieces of material from the time of the crash would have been buried even deeper, making any attempts to detect them magnetically a futile exercise, although we did try bravely, working our way in parallel lines along the zone where the ovoid object had rested. Jose, Bill and I spent the rest of the afternoon searching the area downstream (Figure 18) without any luck in detecting anything of interest along the wash.

At a local Thai restaurant where we relaxed and spoke of Dr. Hynek and of other crash cases over dinner, Paola reached into her bag and extracted a fractured plate of rough metal she had been given on her first trip to the site, before the gully had formed. It had been found (along with two other pieces) by another experienced investigator who'd been intrigued by the story, James C. Fox. Inspecting it, I found it had a hole with fillets for a screw, and a quick test showed me it turned out to be of a standard diameter, clearly pointing to a fragment of an ordinary plate, likely broken off from the Marconi tower when it was hit. In fact I bought the exact screw that fit nicely in the hole, and showed it to my disappointed friend. Another trail leading nowhere.

Back in Silicon Valley, it took a few hours for specialists to confirm the identification of the plants I had gathered with some trepidation in spite of the gloves I was wearing, and brought back in sealed plastic bags.

"The Park Rangers were right, it's plain cocklebur indeed," said the scientist who inspected the plants, pushing back his microscope, "but those are only poisonous as seedlings, at the cotyledon stage. If cattle eat the newly-sprouted plants, then it is toxic to them. Basically, the 'seed' is still toxic when it first hatches. When the first real leaves grow from between the cotyledons, the toxin is gone." He laughed, adding: "At least, that's what the literature says."

"Why would anyone plant such a nasty thing in an area where you raise cattle?" I asked.

"Beats me," he answered. "That's pretty evil. If they feed on them at the wrong time, the cows will sicken and possibly die."

Who said that studying UFOs wouldn't help you perfect your education? Every site survey I have done, every instance of climbing over hills and canyons to reach a landing site, has taught me something special about the precautions you must take, the need for references, the details of good orientation, and especially the fine sense of discrimination you must exercise, between what is really important and will turn out to be irrelevant in the end. I am still learning.

∧ ∧ ∧

What else had gone wrong? I wondered. And most importantly, what was missing?

Most scientists interested in the UFO phenomenon focus on the physics and the primary data about the aerodynamics, like good students at the time of an exam. If they could make sense of those parameters, they might be able to build similar flying devices; they could patent them; perhaps that would give the US a military advantage over an adversary, or a hint for a revolutionary space-faring enterprise.

As a scientist, you start from what's most relevant, mainly what can be measured or estimated: mass, size, shape, altitude, speed.

If you're lucky, the witness gives you a reference to points in the landscape, a trace on the ground, a correlation with a physical effect: vibrating metal signs, burned vegetation, waves in the water, a demagnetized battery. You ask about the light, its intensity and colors, its correlation with acceleration.

The computer files compiled by many private groups contain such information. So do the records kept by the military. Yet any good scientist will tell you that, in order to be useful, such data needs to be correlated or calibrated against something more specific: a known effect, a reference signal. It's a practical requirement in most situations, because there are rational boundaries to human observation. So you ask about the exact time, about environmental facts that any witness would be familiar with, the wind, the temperature, the weather, the sounds in the area.

All that will help place the witness back into the scene, and stimulate the brain to better recall details.

A pilot in the cockpit of a jet fighter will not have the same sense of the situation as an operator in front of a radar screen or a sentinel frozen on the spot by a light from above, which he doesn't understand.

This is where the UFO Phenomenon proves to be most intractable, unfortunately, rendering most typical databases of limited utility, no matter how massive.

Then it dawns on you that the analysis of such an event isn't like a straightforward physics exam, where you can line up a few equations, discover patterns among the numbers, make hypotheses, plug in the known figures and just do the math.

The wise old recommendation to "shut up and calculate" doesn't quite apply any more.

∧ ∧ ∧

We had done a fair job of reconstructing the trajectory, inspecting the site, analyzing the plants, sequencing the days, the times, the duration, the constraints of the weather. We had, in fact, calculated what could be calculated. Yet we couldn't quite make enough sense of it all to begin proposing reasonable scenarios about the craft: where it came from, how and why it crashed, and what its sudden intrusion at that historic moment actually meant.

The obstacles became obvious when we reviewed the results of our two site visits, our good talks with Mr. Padilla, and the rich observations others had noted before us.

What was still missing in our analyses had to do with the human dimension, the social background, the cultures at the site, the languages spoken, even the variable meaning of words. Over half a century had passed. America had changed deeply. It wasn't just the psychology of the witnesses that was at stake, it was their pre-established background, the diversity of the population around them, its traditions and beliefs. And such variables have rarely been taken into account in the analyses of UFO observations or in the compilation of sighting catalogs.

When the witnesses are members of the military, there is an existing, fixed process for interrogation and processing, and it doesn't start with anthropology.

Even when the case happens among civilians and is reported to the police, the procedure begins and ends with "Just the facts, Ma'am".Some subtle questions never come up. Nobody cares if the witness has been terrified by recurring dreams ever since, or has experienced other strange incidents, years later. Investigators are human and they behave according to patterns they've long practiced—patterns that work well in conventional events like car accidents, or a case of stolen goods.

Take the very first observation by the two kids: plenty of hints should have called our attention to the complexity of their testimony.

When they first witnessed the disarray around the crashed craft, the gaping hole in the side of the object and the three creatures aboard, and when they heard the cries compared to those of a rabbit being killed, Jose and Reme remained frozen there *for an hour and a half*, in total contradiction to their nature as active, busy country kids.

They were not physically paralyzed, yet some force glued them to the spot in fascination for a long time. What went through their eyes, their ears and their brains? Why did they feel the creatures were "feeding things into their minds?"

When they rode home that night, with Reme crying in silence, both had images in their heads that they couldn't relate to their current lives, visions that transcended their experience: Things in the sky, catastrophes revealed beyond the mind of a child, confusion.

There was also the sense that the sounds of nature had stopped: no singing birds, no rustling of leaves: a characteristic feature of many accounts reported later, in so many cases of close encounters across the US and other countries.

And for several years after that, the nightmares, the realization that another level of existence impinged on their ordinary American lives; until the true dimensions of the events at Trinity became obvious.

∧ ∧ ∧

CHAPTER EIGHT

THE SECRETS ARE EXPOSED (2003)

Knowledge of the San Antonio crash should have died without notice, except for a chance occurrence, an unrelated Internet search in genealogy archives that reconnected the two prime witnesses after their many years of silence. The Army Air Force had long drawn a curtain around the episode; all the young soldiers had gone on with their civilian lives.

In 1947 the Army Air Force itself vanished into the limbo of military bureaucracy and whoever got the brief at the top security office of the newly-created Energy Department and at the Atomic Energy Commission clearly understood that her job was to bury the folder.

On the academic side, too, the White Sands scholars had moved on, ignoring the devastating damage radioactivity would pose for a long time to people in that area of New Mexico, erasing local anecdotes about Project Manhattan.

Even the political overtones had been expunged.

As Reme Baca had vividly recalled, if you talked too much about the affairs of Ground Zero, or if you happened to remember too many details, the government might send you on a long restful cure in "Las Vegas," the local nickname for the nearby insane asylum. The treatment might not be as harsh as it was for people who talked too much in communist Russia, but there were notable parallels. Everybody was aware that World War Two had just been extended into a Cold War, which would last longer than the shooting war and was very nasty in its own way.

By then, historians were focused on the bomb itself, understandably fascinated by all the new physics they had missed in college. Writers applied their talent to analyses of the closing days of the war with Japan. They had no interest in what unusual phenomena or creatures a couple of country kids might have seen one month after the bomb.

It's only in 2003 that the two friends, Remigio and Jose, reunited through a smart genealogy search over the Internet, compared notes again and recalled the details of their experience.

They realized the significance of what they had seen and decided the time had come to tell their story to a journalist, a former classmate at San Antonio Grade School.

His name was Ben Moffett.

They trusted him because he was the only one they knew who was familiar with the culture and the ambience of the area, its landscape and its people. In 2003, he published the very first article ever written about the 1945 crash in a local paper. But it would be another six years (precisely until May 2009) before Paola Harris learned about it in Rome through a newspaper article. She told me how it all happened:

After she resettled permanently in the United States, she'd contacted Remigio Baca and his wife Virginia, and in July 2010 she travelled to Gig Harbor, Washington, to interview them in person. There, she saw the "bracket" Jose had removed from the craft, and Reme showed her a notebook with an analysis done at Los Alamos by his brother in law.

The analysis in question showed the piece to be made of a "plain old Aluminum alloy."

The details came out in Reme's later interview with Paola, when he told her:

"When I came back from California and retired in (the State of) Washington, and then when I talked to Jose, Jose and I were communicating back and forth on genealogy. And then Jose reached out, we started getting together and gathering data on what we'd discovered there. Jose was sending me information that his son (Sammy) had found, and I was going through it."

Chapter Eight - The Secrets are Exposed (2003)

"There was a newspaper article that Ben Moffett, a friend of ours whom we had gone to school with, had written in one of the Socorro newspapers a couple of years before. **(20)** This included a visit to the San Antonio Elementary School where he talked to the teachers, and so they were briefing him on present teaching methods..."

By then the war and its aftermath were forgotten. People had moved on. All the excitement was not about atomic secrets any more, but about computers and the Internet. Even the Apollo era and the moon landings were old news. It would take a while, even for that local journalist, to take a serious interest in the crash at San Antonito.

^ ^ ^

The next step was to reconstruct the exact sequence through which the incident came to the light of day. That was obviously important, so Paola pursued the matter in detail, asking Reme Baca how Ben Moffett's newspaper article dealt with the history of the region.

Reme: They were trying to update him on the use of computers (...) So in the process of doing that, he named some of the students that were in his class in the article, and omitted Jose and I...We thought he might be surprised that we were still around, so I gave a call.

Paola: You noticed the lack of your names in the article he'd written about his classmates, and you kind of felt like he didn't remember you...

Reme: So I called him up and as soon as I said, "This is Reme", he said, "Oh, oh, you know, I couldn't remember your name. That's why I didn't..." and he apologized for that. We got to talking and he was retired, and was working for the paper (...)

"Well, you're retired. You must be doing something."

I said, "I'm working on genealogy."

"What else are you doing?"

"I'm trying to research this object we discovered on the Padilla ranch, when we were kids..." His response was that "he wasn't into UFOs." But a few weeks later, Ben called me and informed that he wanted to do the story, and that's how this all began.

Paola: Okay, in 2002. So, when did Moffett write the story?

Reme: In 2003, I believe.

Paola: Because you know, I read it in Italy, as a story. I was blown away because it hit me. I couldn't understand how people hadn't researched it.

The fact is that nothing more would have happened, even after the newspaper story, and after Reme Baca and Jose Padilla were in communication again, if Paola hadn't decided to pursue what was, to her, a very important piece of history we had all missed.

Reme tried to find information on the Internet, and came back empty-handed. So he talked to some ufologists, and they ignored him because they thought, "If it was so big, why didn't it come out?" He seemed to come out of nowhere, making this story up so he could get attention. As Paola put it: "If this were true, it would have been as big as Roswell!"

Reme picked up on that remark: "Even the people that investigated Roswell, I talked to them…And the other one, this doctor, whatever his name is, the atomic…the guy that…"

Paola: Oh, you're talking about Stanton Friedman. **(21)**
Reme: I eventually talked about ten minutes with him. And he says, *"You know, when I was investigating Roswell there was a phone call, a guy that called, and he was telling me about a crash in San Antonio. But it wasn't important at the time, so I didn't really have any reason to follow up on it."*
Paola: So the progress of your story is that you made a little attempt to talk to people to be taken seriously, because part of you really wanted to talk about this, but you found that people shut you down, because first of all -- I'll tell you why they shut you down: *The story is too big*. It's a huge story. It's not just, "I had a little sighting and there was this…"

> This is huge. It's very, very big. It could be bigger than Roswell if it would have more witnesses because you have the pieces, the artifacts to prove it. So if you had to guess, how many witnesses do you think there could be, that would talk? Okay, there is Jose, you…

Reme: We've gone through the whole list.
Paola: So after Apodaca died… What about Apodaca's family?
Paola: They're gone. Age catches up with you!
Paola: Who else?
Reme: Well, first of all, the only witnesses, the only people that were involved, was Jose's dad, myself, and Jose… There were a couple

of people... one of the people that was there that married Jose's cousin...
Paola: Is that person still...?
Reme: No, they're gone.
Paola: And even the sheepherder...
Reme: They were old, then...
> Paola: So what does that tell you then? That you have the responsibility...that you're the only ones left. I mean, there's nobody else. Otherwise this goes untapped. The history of this phenomenon which we're trying to archive as history, you know. There's nobody else.

^ ^ ^

*There's nobody else...*so we now have the challenge to study one of the richest cases of close encounters, and the earliest one in post-war history, a case that *the witnesses kept to themselves from 1945 to 2003.*

As we're about to see, there were good reasons for their silence. It came in part from the simple fact that as they grew up, Jose and Reme went their separate ways and moved into different occupations—one in Washington State, the other in California. But there were other, more scary reasons. The atomic bombs had put an end to the shooting war, but the Cold War was just beginning. Secrecy and paranoia, if anything, were becoming even more imperative and intense.

Everybody was scared. Because everybody was under suspicion.

In 1954, Dr. Oppenheimer himself was dragged before a four-week long Atomic Energy Commission Security Hearing that revoked his clearance (by a two-to-one decision) and left him a broken man, although it was acknowledged that he had never betrayed any secrets. Roy Glauber, interviewed in 2013, recalled seeing him briefly after the Hearing: "His hair had been turning grey, what there was of it, and he looked worn and exhausted." **(22)**

Today a detailed write-up in Wikipedia notes: "The loss of his security clearance ended Oppenheimer's role in government and policy. He became an academic exile, cut off from his former career and the world he had helped to create. The reputations of those who had testified against Oppenheimer were tarnished as well, and Oppenheimer's reputation was later partly rehabilitated by Presidents John F. Kennedy and Lyndon B. Johnson. The brief period when scientists were viewed as a "public policy priesthood" ended, and thereafter would serve the state only to offer narrow scientific opinions."

Scientists working in government were on notice that dissent was no longer tolerated.

This was hardly an environment where genuine, responsible witnesses of extraordinary phenomena, so close to a strategic military zone, would decide to come forward. And it is not an environment where one can do good science, either.

As for the full story, it can only be reconstructed today, as we are doing, step by careful step, in this book.

The next level in the testimony of Reme Baca would deal with the actual scene and his recollection of the details. It turned out, to Paola's delight, that his memory was quite clear.

∧ ∧ ∧

The general outline of the story has been presented in earlier chapters, but it cannot be fully grasped without the smaller details, the local constraints, the finer motivations for people to do certain things; or not do them. And one can only know the actual facts from direct interview of the witnesses, which may align neatly in some areas and differ in others, based on their location and their perceptions at the time.

Reme Baca was gracious in his willingness to recall that experience of 1945, when he was a child of seven, confronted with an event that has mystified the scientists, the military, and those who have hidden the data since the end of World War Two. His wife Virginia ("Ginny"), who had often heard the story, supported his testimony, telling Paola: "The area now, where that ship hit the ground, it slid a long ways. And so, it left bruises in the land."

Paola: And that's still there?
Reme: Yes. I showed you the pictures.
Virginia: And also that tower…
Paola: Is the tower still there? **(23)**
Virginia: No, it hit it. They figured it hit it.
Paola: Was it a radar tower?

Chapter Eight - The Secrets are Exposed (2003)

Reme: It was a radio or radar, we don't know what it was. It could have been radio... It was an about 50 foot tower and it had what they called a wind charger, a little propeller, and it hit that. And we figured that the sound that we heard was, it got hit between the lightning, because lightning probably would hit that tower. The tower probably had a lightning arrestor, so it hits it and it's grounded, goes into the ground. But the craft probably was close to that tower and lightning went through the craft and that's what caused it.

Paola: The malfunction? The whole …

Reme: ...*and then they were able to maneuver that craft so it wouldn't tumble*, because it should have. It should have, it should have tumbled.

Paola: In other words, it slid and then made a right-hand turn…

Reme: …to stop against a big wall of dirt. We did everything that we possibly could, because we were young kids, going through every record, talking to just about anybody that would have known anything, other than us…it was done in such a way at that time that it didn't … it really didn't bring any attention.

Fig. 19: On the hillside: reconstructing the object's trajectory.

At this point in her investigation, Paola needs to take the narrative from the point where the craft hit the tower to the actual recovery of the metal bracket, so that the exact details can be scrutinized one more time. There is no room for ambiguity or "guessing" in such research: You just get the story, and then you come back and you do it again, from another angle.

Paola: One other thing: Tell me about the café, the watering hole that Oppenheimer…I need to know because I'm fascinated with Oppenheimer. If you read the life of Robert Oppenheimer, right until his death, he had all kinds of problems. …He went against nuclear, his top secret clearance was taken away, and Kennedy gave it back to him (…) He went through hell, because of that situation. So, the watering hole was called what?
Reme: The Owl Bar and Café.
Paola: And you said it had a fountain, an ice cream fountain?
Reme: Fountain, that's what they called them: a "fountain."
Paola: Yes, I remember. Because that's where they made…I lived in the days of the fountain!
Reme: They had a jukebox and a pinball machine.
Paola: A jukebox, a pinball machine, and a bar. You could have liquor…
Reme: There was a bar, and they made hamburgers…
Virginia: They still do.
Reme: Yes, they still do. And chili. They made hamburgers with chili, "green chili burgers" they called them.
Paola: And back in 1945 that was open. And Oppenheimer used to stop there?
Reme: Oh, a lot of them scientists…
Virginia: It was the only place to stop.
Reme: 120 miles from Los Alamos to Trinity… (Note)
Paola: Did you ever see Oppenheimer?
Reme: I saw a guy that looked like him. I didn't know who Oppenheimer was at the time.
Paola: He was a very thin man, very thin.
Reme: He wore a little hat.
Paola: Okay. And so you would see these people come in and out of there…
Reme: …and hear some of the stuff they were talking about. And I was not all that well advanced in English and …

Note: The actual distance is more like 150 miles, according to my maps.

Chapter Eight - The Secrets are Exposed (2003)

Paola: How would they get there? With cars?

Reme: An entourage.

Paola: An entourage of cars. And that's where those guys stopped at night... that day? Was it day or night that you jumped on the... when you got the piece (*from the crashed object*)?

Reme: *It was at night.*

Paola: *It was at night,* and that's when all those young guys stopped and the flatbed truck was there, so you guys could jump on and pull that thing off the panel?

Reme: They'd leave the site and go there to lunch, and also ...

Paola: *But this was at night*, and they couldn't see you jump on...?

Reme: No, we're far away from the Owl Bar.

Paola: Where they left the truck was far from the Owl Bar and Café?

Reme: Oh yeah, *it was back at the crash site.*

Paola: So you did it at the crash site, while they were at the Owl Bar and Café. I thought it was like they parked it outside.

Virginia: No, they were still at the crash site. They would leave at night.

Paola: They would leave at night. So that was the last night, and you decided to go and jump and get that piece off there, off a panel...

∧ ∧ ∧

We should pause again here, to note a discrepancy between Reme Baca's recollection of that evening's events and the later testimony of Jose Padilla, who told Paola that the sun was still above the horizon when he climbed into the craft. In fact it was that set of circumstances that allowed him to obtain a clear description of its interior.

The conversation also went on to note the proximity of the San Antonito crash to the extraordinary observation, in April 1964, by Highway patrolman Lonnie Zamora, of a similar ovoid craft that landed just outside Socorro and left deep imprints in the soil, only about ten miles to the north, with the involvement of the same families. **(24)**

Paola remarked: "You know, I've heard this so many times, from so many people. Lonnie Zamora's wife is Jose Padilla's cousin, that's an amazing coincidence."

Reme: Remember, there's only six families…

Paola: The Zamora case is a very major case because he watched the whole thing, you know, in that area. And what was that? Ten miles away. So it was a hotbed of situations, this place. Not only the Zamora case, but this case….

Reme: … There were other members related to the Padillas and Bacas that had ranches there, that also knew the Fosters from the Roswell story, and they communicated with them a lot. At that time, you know, the ranches weren't very close. They were far away so it was a jaunt to get from one to the other. We had people that lived there that worked at Roswell and came on leave (…) So in the evenings some of the people, because it was a little town, not much to do, some of the guys would come in and they'd talk about Roswell.

One of the things that was very clear to us was that the people that knew something, or thought they knew something, about Roswell, would end up in what they called "Las Vegas Hospital…" It was a mental institution in Las Vegas, New Mexico. …

We also know, and a lot of people probably still do, but after our crash there, after that crash… in 1945, there appeared to be a lot of sightings…they called them ghosts.

I recall one guy that was at a dance there, and he was on his way home and he started running, and couldn't see what he was running from. And he said there were some creatures that were after him.

He thought they were devils or something. So there was a lot of apparitions, I guess they call them ghosts. So, a lot of things were taking place after that time.

And there was people coming in that didn't belong there. In a small town with about fifty people

Paola: Intelligence types? "Men in Black?" I don't know what you want to call them?

Reme: No, they weren't "Men in Black," just out of place. You don't wear a white shirt in San Antonio, unless you go to a funeral or a wedding. You wear a suit, for that matter.

Paola: What years are we talking here?

Reme: We're talking about 1947-1948.

Paola: After Roswell, okay.

Chapter Eight - The Secrets are Exposed (2003)

Reme: And so people pretty well got the message...*keep your mouth shut.* And they were used to that, the Manhattan Project happening, keeping their mouth shut, and they did. But some of them pushed the limit, and so they got caught up in that. They didn't question our ability to keep our mouth shut. *Whenever Jose and I talked, we didn't talk at school.* We'd go over to his ranch and talk, out there in the middle of the hills, and talk about stuff when nobody else was around. We were very careful. Because we knew what the consequences might be.... We're not into this stuff about Roswell.

^ ^ ^

In the final part of her 2010 interview, Paola goes back to the topic of the publication of the case: How would the witnesses want to see that process happen, after so much time?

Paola: And your story should come out how? How can we best present your story to the general public? What did you have in mind?
Reme: We'd like to have a dig, you know, go over and dig. Either call it a trench or whatever, where some of the soldiers threw some of the pieces, and that's been covered over, over time, and we'd like to dig that up and see if there's any pieces remaining. (...)
Paola: The soldiers, why were they throwing them in there? They didn't want to carry them?
Reme: I believe so. You have to, I guess, give those soldiers credit. We discovered this craft exactly 30 days after the bomb test at Trinity. They were restricted to base camp at Stallion Site for 90 days. They couldn't go out of there nowhere, and talk to anybody, or have a drink of pop or anything else. They were restricted to that area until the bomb test was done. And then after the bomb test was done, they released them so they could have a little liberty and so forth. And then this crash took place and they were working again, doing the recovery! These were young men that took every opportunity to socialize with the people there, it was probably the first time they had been able to do that in 90 days.
Paola: So they really didn't.... it was not a methodical, paranoid picking up of pieces. In other words, "let's get this stuff cleaned up and go have some ice cream or some beer, and who cares whether any part gets left behind?"

Reme: I don't believe there was a formal alien craft recovery-training course! We were at war.
Paola: And you even said they weren't even wearing biological suits, they were not available.
Reme: They weren't aware that this was an Alien ship, and we weren't either.

Now Paola reminds Reme that he's lived with that secret for over 60 years... He responds:

Reme: That's a long time. Well, not really lived with it for 60 years, because at a certain point in time I could see that it really wasn't possible to do the things that we wanted to do, you know, while I was here in Washington (State) going on with my life and such, raising a family and so forth. Jose was doing the same thing in California. So there was a point in time that we were not concentrating on this. Perhaps we weren't ready for it.

^ ^ ^

In 2017, Paola had completed that first cycle of research by going back to Jose Padilla. Again, the subject was a possible return to the site in the hope of resolving some of the details, and perhaps discovering more hard evidence of the object:

"Whenever you're ready to go, we'll go," she told him. "I don't know what Reme and his wife are doing either. I don't know what Ginny and Reme are planning, but the end of September sounds good to me, because it's cooler and if you would plan a definite time (...) maybe you'll find some more pieces, because of the water that erodes the ground, there might be pieces there. There's been a lot of looking. Just people...that are walking around over there."

Mr. Padilla reminded her that he was "born and raised there."

Paola: Oh, I know you were. I know you know what happened, but do you think...they're not tourists, right? They're just curiosity seekers?
Jose: Curiosity seekers, and there's a lot of rifle shooters doing target practice there.
Paola (*laughing*): Well, that's not going to be fun!

Chapter Eight - The Secrets are Exposed (2003)

Jose: That's just target practice; they set up targets, like bottles, and cans. I can understand city folks would have a hard time understanding this.

Paola: You're the one that actually took that crowbar or whatever you took, and pulled that piece off, right?

Jose: Right. I'm the one that used the cheater bar to jerk the piece off the panel.

Paola: And you saw the panel it was on?

Jose: Yes. You can see that, that it's clean. The pieces that are there now....

Paola: This would be 62 years later. Yes, you know, it's a lot of years!

Jose: The piece was taken care of. Reme kept that. It was his *tesoro*. His treasure...

Paola: So, is there any more of that anywhere?

Jose: I don't think so.

∧ ∧ ∧

With the new plans agreed upon, but the exact schedule left undefined, the conversation returned to the subject of the beings that had emerged from the craft after the crash:

Fig. 20. The Silumin bracket (photo: Paola Harris, 2019)

Paola: Can I just ask you, really quickly, what do you remember the creatures looked like?

Jose: Well, we were little kids then. I was only about three feet tall, they were a little shorter than me. Kind of light grey.

Paola: Do you remember if they looked directly at you?

Jose: I don't know whether they were looking directly, I know they were running back and forth, squealing...

Paola: That's actually a good word, because Reme was also saying they seemed to be "squealing in pain."

Jose: I think they were hurt.

Paola: Okay.

Jose: Because I wanted to go in there to help and...

Paola: But you had three days' worth of going back and forth there.

Jose: We had a week, when they cleaned it up. We used to take our horses on that ridge.

Paola: Did they see you? They didn't see you, did they?

Jose: We were pretty sneaky. I know my territory!

Paola: You were pretty sneaky. So you saw the soldiers throwing stuff in that ditch, too?

Jose: Yes, whatever they didn't want to pick up, I guess, those pieces were too many.

∧ ∧ ∧

When Paola came back to the idea of a return to the site, the next question was to find a convenient date for both Reme and Jose. Jose went along with the idea: "There'll be no problem," he said. "It's not the target shooters, the ground needs to be warm."

Paola (laughing): Okay, it's not about the target shooters, then, that we're worried about...

Jose: No, I don't think so. Because things have quieted down quite a bit since then, so....

With Jose's basic information in hand, the conversation returned to Reme's experiences as a sequel to the sighting itself. We have to keep in mind that Reme, two years younger than Jose, was crying at the site and had a hard time dealing with the experience as the two children rode back home in the dark.

Chapter Eight - The Secrets are Exposed (2003)

Paola: Have you had any weird experiences, after your sighting? Any weird dreams, anything that you think is related to your sighting the craft and being so close to it? In your dreams, what is the space-oriented stuff, other than the fact that you saw a kind of catastrophe from a skyscraper, you saw people falling, and all that kind of thing....

Reme: Well, some of it was where I was on the ship, (*note: in the dreams*) and there were others in the same ship, but further down. (...)

Paola: Did you ever see the beings? Were they the same kind of people?

Reme: No, I didn't see the beings. And it didn't make no rhyme nor reason.

Paola: Well, did you see (it) while you were sleeping, or just daydreaming it?

Reme: I wasn't daydreaming it. I was in bed. There were some freaky things that I saw. After a while, it's a drag, because you don't really feel good about it.

∧ ∧ ∧

PART THREE

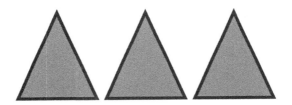

THE CASE OF THE SILUMIN BRACKET

The problem with UFOs is that somebody is creating Reality for us, and we don't know who it is…Could there be a consciousness that is shifting the mean point of the human system, altering our statistics in subtle ways?

All it would take is to create an experience that changes a few individuals.

Professor Robert Jahn
Dean, School of Engineering
Princeton University

CHAPTER NINE

THE INVESTIGATION

"After a while, it's a drag, because you don't really feel good about it," Remigio Baca had told Paola in 2010. And that's how I felt myself in 2019, flying home from New Mexico after my second trip to the site. If anything, we had more questions than we did at the beginning.

When I spoke to her again, in early March 2020, Paola was living in Colorado and the world had entered "shelter-at-home" mode because of the coronavirus pandemic. I suggested to her we needed to take the opportunity of the sequestration to re-examine everything we knew, and "take it from the top" again. Except that now we had to do everything at a distance, through the telephone, the Internet, and the dubious wonder of a leaky, insecure but surprisingly convenient piece of software called "Zoom."

Reme Baca had passed away by then (**25**), but based on our investigations at the site we were able to carefully re-interview Jose Padilla, now aged eighty-two, who was no longer worried about liberating the infamous "bracket" from the US Army since he was well beyond the reach of the long arm of the law. Given his age, he was eager to help us understand and document the events of 1945, once and for all. As for myself, having just turned eighty, I was eager to listen to him, and learn from him, because what he was saying could help bring clarity and substance to a mystery that had intrigued me, professionally and personally, for most of my life.

We had to begin again, I told Paola, starting from the craft itself. We had enough of all the esoteric speculation: we needed to know actual dimensions, weight, materials, and a general model of the outside and the inside. So we drew up a new list of questions and she tapped Jose's excellent memory, once more.

∧ ∧ ∧

The first set of questions, as we reviewed the earlier transcripts for another level of precision, had to do with the metal and the "scraps" that had littered the ground following the crash.

Paola: OK now, what were those soldiers picking up, if there was nothing there? And there was a panel: What in the world were they picking up?
Jose: Pieces of… like, crashed metal… aluminum: when it breaks, it doesn't break in parts, it disintegrates. There were pieces of disintegrated metal that they were pushing into that hole there.
Paola: Did you see any of them pick anything up?
Jose: They kind of kicked it. See, they were supposed to pick this up, and bundle it and put it into a box or something, to take it in. They didn't. They were too lazy to work. They just went and pushed it into a little crevice, you know, and put dirt on top of it.
Paola: And did you guys… did you do that too?
Jose: No, no we didn't.
Paola: Well then, what…
Jose: After that happened, the following day and a half, when we came back with Eddie Apodaca, the policeman, someone had cleaned up, like with a rake, and there were no pieces on the ground.
Paola: OK, so the pieces that the soldiers were pushing with their foot, would you say they were little pieces, or were there any big pieces?
Jose: No, no, there were just pieces that were…, heck, you can just pick them up with your hand, you know. Disintegrated metal.
Paola: Disintegrated metal. I can't picture what that's like, but…
Jose: It's pieces of metal like, OK, cast iron, let's talk about cast iron now.

Chapter Nine - The Investigation

Cast iron you cannot break. It will disintegrate. If you hit it with a hammer it will fall into pieces.

Paola: So that's what the debris was, but if it was a panel and the panel was not broken, what part was broken?

Jose: It must have been....I will say, you know, it was a part where it snapped out of...out of the object. The pieces of metal there, that these guys were shoveling into the crevice, were probably pieces of a hinge or a bracket that was holding that panel together.

Paola: You know Jose, when you say that, it sounds like this was manufactured. Yet you didn't see any screws or anything, it was all smooth. It looked like you're talking about something that might have been manufactured. But even the outside of the craft, was that all smooth?

Jose: Yeah, it was smooth.

Paola: And, you said it was grey. Was it shining in the sun?

Jose: It was a grey, a grey metal that ... might have probably been through smoke, or this was heat, while it was floating down or something.

Paola: In other words, when it crashed, it went through a heat situation.

Jose: Well, no. There was some scratches where it was, skidding on the ground.

Paola: There were scratches?

Jose: Yeah, on the outside panel. But it was a smooth area, you know.

Paola: What was the weather like when all of this happened?

Jose: It was wet. It was raining, but then the rain quit. But when that...I tell you, that thing was so hot when it hit the ground, that it started a fire! And the boom...

Paola: ....when you heard the boom or the bang, and you thought it was the bomb again, was there a lightning and thunderstorm?

Jose: No, this was after.

Paola: Well, was there ever a lightning and thunderstorm?

Jose: There was a lightning and thunderstorm, but when that "sonic boom" happened, I believe that's when that object hit the tower.

Paola: Could the condition of the thunder and lightning storm affect that thing, do you think?

Jose: Well...my...(*chuckles*), my opinion is that...that "lightning" grounded that object, whatever it was. And it brought it down, and it hit the tower.

Paola: The lightning grounded the craft, it grounded it, it hit the craft, and then it went and took out the tower, and then it crashed?

Jose: Right: it crashed down into that ground there, where it skidded all the way ... to a stop.

Paola: Well, Jacques may have more questions for you, because he's trying to recreate it all.

Indeed I was. From experience with the field I knew the story we were piecing together from the sometimes divergent accounts of the two main witnesses, would be challenged, distorted in a dozen podcasts and buried in false or deliberately misleading information by naïve—or not-so-naïve—individuals who tend to show up around such investigations. But there would also be serious inquiries from genuinely interested scientists.

We needed to do the best job we could, pinning down the physical characteristics of the object itself, *as a manufactured craft*. Any analysts following us or reviewing our work would want to know dimensions and materials – two areas that Mr. Padilla had very competently covered in our interviews – but also weight, density and other features.

The object was long gone, so the best we could do was an estimate. But that was within our reach, based on the parameters of the tools used by the Army to retrieve it, namely the crane to lift it and the truck to carry it away: and this led us to generate a whole new list of questions.

∧ ∧ ∧

Paola: What we need to know is a couple of things. When you walked up to the flatbed truck, how long do you think the flatbed truck was?

Jose: The trailer? *Forty* feet long.

Paola: How long did you think the craft was? It was tilted on the side, wasn't it? On one side?

Jose: Let's start over again. The truck... The flatbed trailer was forty feet long; the object was thirty feet by fifteen feet.

Paola: Thirty by fifteen, OK. When you went in, you went under the tarp. Was it dark when you went under the tarp?

Jose: Yes it was.

Paola: It was dark, so how could you see anything?

Jose: Well, when the tarp was full up, through the hole, the sun was shining, because the sun was coming in from the West.

Chapter Nine - The Investigation

Paola: So, when you pulled the tarp through that hole, co[uld you see what] was inside?
Jose: It was cleared.

    Jose was telling us that the soldiers, by then, had removed a[ll the] objects that might have been located inside the craft. On site, the m[en] was now ready to take the object to an Army base or a laboratory somewh[ere,] securely attached on top of the trailer. We were making progress, buildi[ng] up an impression of the physical circumstances. But many questions remained.

Paola: We need to know, did you touch the wall? Did it feel like metal, did it feel like plastic? What did it feel like?
Jose: *The floor was metal, the siding was all metal, too.*
Paola: What did it feel like, was it like steel? Did you touch the wall?
Jose: It was metal paneling.
Paola: Oh, it was *paneling*? You mean, there was separations?
Jose: Metal paneling on the inside, all the way around. See, the object was an oval shape.
Paola: Why did you say "panel?"
Jose: The paneling, like on inside the wall: it was paneling: it was metal.
Paola: Did you put your hands on the wall?
Jose: Yes I did.
Paola: And you couldn't see any furniture? Or anything inside?
Jose: It was clean, as if someone had been there, and it had cleaned everything up, except for that piece that you got (*Note: the artifact shown on figure 20*), and the paneling...
Paola: That was the only... the only thing that you saw? With that panel on the wall?
Jose: Yes; and the little dome, up on top.

∧ ∧ ∧

    You often read witness reports about saucer-shaped objects with a transparent or translucent dome of top. I had observed such an object myself in Pontoise in 1955, with two other witnesses (one of them half a mile from me, equipped with binoculars). But the craft Jose was describing to us was very different: it was not a disk at all; it was egg-shaped, or

...nall glass or plastic bubble, not on top but ...drawings (Fig. 21).

...aola, when she asked: "And the dome, ...de of?"

... a plastic.
... a plastic. Was it big enough to fit one or two

...this thing was about 20 inch or 30 inches wide, and an oval shape, and you could see through it.
Paola: Did you go up on that oval, and see? Could you go up there?
Jose: No. I couldn't go up, because it was tilted, but *I could see through it* from the floor.
Paola: Oh, that's interesting! You could see through it from the floor. Could a man fit up there?
Jose: If there was a ladder, or some kind of object, that would hold a man up there.
Paola: So, if that was a little man…Could a little man go up there and see?
Jose: Yes they could, because he would be smaller than us. It was like a watch tower or something.
Paola: But could you go up there at your age, or was it too small for you? (*as a boy*)
Jose: No, I could have gone up there if I'd had a ladder and a kind of handle I could hold onto…
Paola: So, the inside of the ship, you could see because the Sun was coming in?
Jose: Right.
Paola: So, you remember that very clearly?
Jose (*energetically*): Oh, yeah!

^ ^ ^

We still had two questions on our list, about the final impression Jose drew from entering the vehicle, and his recollection of the diminutive occupants. Those were items our sequential interrogations had covered before, but there were some discrepancies between Jose and Reme that reflected the depth of the missing information, and the natural differences in psychological reactions to a traumatic event.

Chapter Nine - The Investigation

Paola: Then you put the tarp back on. OK, now, the hole? Was the hole perfectly round, or was it jagged? Where it popped out, was it jagged?

Jose: It seems to me like a whole panel popped out...

Paola (*surprised*): Oh, so, go ahead...

Jose: It was like a whole panel, you know, when it hit the tower, it popped out. It was about four feet wide by almost nine feet tall, high. The ceiling was thirteen feet, inside.

Paola: The ceiling was thirteen feet, inside. So when it hit the tower, it wasn't a jagged hole: What you saw was a panel that flipped out.

Jose: It looked like a panel just broke right up.

Paola: It was part of the ship, but it was a panel, it wasn't all jagged or exploded.

Jose: No, it looked perfect, except for the pieces that were, that fell on the ground, you know. But the panel was a big panel that blew out of it.

Paola: Did you see any of the material that folds up, like the aluminum, that Reme had, inside?

Jose: No, that thing was cleaned out, except for that mechanism that I pulled out (*the bracket*).

∧ ∧ ∧

Paola: Now I'm going to ask you about the little creatures. Those little creatures, what color did they seem to be, when you looked through the binoculars?

Jose: Either they had a uniform on, or their skin was like... a light grey.

Paola: OK, you think they had a uniform on, or the skin was a light grey. And what about their head?

Jose: OK, they were short, and they had long arms that went all the way to their knees.

Paola: And how was the head? Was the head the same size (*as humans*)?

Jose: The head was kind of a pear, you know, a pear shape.

Paola: Oh, pear-shaped head, OK. Was it big, or small, or the same size?

Jose: The head was just a little bigger than the chest.

Paola: So it was bigger than the chest. Did you think the head was too big, when you were watching it? Or did you think the head was normal?

Jose: I think that the head was normal. The only thing that I was figuring out was that, to me, *they looked like a fire ant, standing up.*

Table 1: US ARMY'S UFO RETRIEVAL TIME TABLE: August 1945

Date	T° and Weather	Witnesses	Craft	Occupants	Military
Thu.16	Thunder & Rain 66/92	Jose & Reme look for a newborn calf. Witness Crash and creatures.	Hits radio tower	Three insectoids	Airplane sees 2 "Indian boys"
Fri.17	67/97	Report to authorities Gardening at home Not on site	No data	No data	Pilot Brothy on site?
Sat.18	69/98	1. Trip to site with State Police (Apodaca) + father 2. Return alone that afternoon	Apodaca and Jose's father enter craft: Shocked	All gone! Site has been raked	Army Colonel Turner takes over
Sun.19	69/100	On site? unknown			Army:Balloon! Ask access
Mo.20	! traces 65/98	Boys on site			Widen gate to 27 feet
Tu.21	! traces 65/81	Boys on site			Bring large truck with a Low-boy
Wd.22	61/85	Boys on site			Soldiers bring a Cherry-picker
Th.23	61/88	Boys on site			Build A-frame
Fri.24	62/93	Boys on site			Soldiers lift object
Sat.25	66/93	Boys on site	Bracket retrieved		Guards relax & the truck moves out

Note: NOAA weather data is from Socorro and Holloman AFB.

Paola: That's because of the shape, right? The shape made you say that: A fire ant?

Jose: Yes.

Paola: Could you see their eyes?

Jose: The eyes were kind of a teardrop, long ways.

Paola: So it was like Asian eyes, almost, right? Teardrops, long ways. In other words, would you say part of the eye wrapped around the head?

Jose: Unless they had goggles. It seemed to me either that was a uniform they were wearing, or that was the shape of their body and their eyes.

Paola: So their eyes were teardrop, long ways, you said, no nose, right? Nose and mouth?

Jose: The mouth? Like a small hole, like a baby's mouth: small, and the nose, were like two little holes.

Paola: What about ears?

Jose: The ears were kind of a little... real small. Tiny little ears, and that hole, you know.

Paola: OK. And they didn't have any hair? So it was a bald head, right?

Jose: Right.

Paola: Now, what noises were they making? Or did you feel like they were communicating?

Jose: To me, they seemed like they were injured, and they were making, oh, I would say, like a rabbit that was injured; or a newborn baby that was crying.

Paola: A newborn baby that was crying. And when you looked at them through the binoculars, did you think they knew you were looking at them?

Jose: Ah, NO! Nobody was watching us; or knew we were watching them. If they would have been watching us, I think they would have come over to us. *These people were scared.* And I knew that they were injured. They were excited, I guess. They were sashaying back and forth.

Paola: By sashaying, you mean sliding? And how many did you see?

Jose: There was three of them.

Paola: You saw three of them, OK. And there was no dead one? You didn't see a dead one?

Jose: No.

Paola: You saw three of them. And did Reme ever tell you what he saw?

Jose: No, he never did. I know he looked, he saw through the binoculars. He's seen two men that I'd seen. I told him that they were injured, that's when I decided that I wanted to go over there and help, and he didn't, and that was just it. (Note)

Paola: Did you talk about it when you went back to the ranch? About what strange beings were in there?

Jose: Yes we did.

Paola: And what did you call them? Because you had no reference to anything.

Jose: I just called them *Hombrecitos*. You know, little men.

Paola: Little *Hombrecitos*. So you thought they were men.

Jose: The way they moved, and the way we moved, it seemed they were older than us.

Paola: So you didn't call them little children, you called them *little men*.

Jose: Yeah. To me, it seemed that I was seeing, you know, *midgets*.

Paola: Midgets. When you think about this, I know you are, did it seem weird? Or did it seem normal to you, because you were nine years old?

Jose: In my whole life I have never seen anything like this. And I was surprised...I wanted to go in there and see what was going on. I wish (*inaudible*) I would have gone in there.

But if I had gone in there, I wouldn't be here to tell you!

Paola (*laughing*): I know that! But when you were nine years old, what did you think, *then*? Did you think they were normal? You thought they were men?

Jose: *I thought they were men.*

∧ ∧ ∧

Paola and I carefully transcribed that tape and we both went over it several times, noting the terms that were used, and admiring the precision with which Jose could recall those events of 1945 with the words of a trained policeman, in order to reconstruct the details for us.

Note: Reme Baca had earlier stated: "the object was occupied **by distinctly non-human life forms which were alive** and moving about upon our arrival minutes after the crash."

Chapter Nine - The Investigation

There was much between the lines; there was much we thought was incomplete or unconsciously suppressed; and there was much we just didn't understand.

It was now obvious to us that the information rested in multiple layers, and that the ultimate key to the phenomenon might not be in the physical, engineering layer but in the more subtle levels of the human environment: The evolving structure of a local multi-cultural, multi-racial society of the magnificent American Southwest, and in the unconscious relationship the phenomenon was able to establish with the minds of its chosen witnesses. And those witnesses had just been engulfed in the first large-scale atomic blast in human history, a fact the ufologists had missed altogether.

After a while, the truth dawned on us:

Trinity was not just a "test," although the physicists thought of it as such.

Trinity meant nothing less than the first, full-scale detonation of an atomic bomb over American soil, with full knowledge that part of the American population would be kept in ignorance, and would be exposed.

Was that the key we were missing, to an understanding of the device that crashed at San Antonito? The scientists thought of their bomb as a physics test; the military thought of it as a weapons test. Historians, today, still use that word: "test." But that is only a first approximation, a gross understatement.

There was nothing understated about the six kilograms of plutonium inside the Gadget.

It was a full-scale plutonium bomb. It was the twin of the bomb that pulverized Nagasaki.

And it did the work of a full-scale plutonium bomb.

∧ ∧ ∧

White Sands may well be a desert, but the 1940 census listed 38,000 people living within sixty miles of Trinity. In their 1991 political science book *Trinity's Children* (**26**) Bartimus and McCartney mention the beauty of the high desert and mountains on either side of Interstate 25 as it winds north from Las Cruces, New Mexico, all the way to Buffalo, Wyoming. But reviewers of their book have been struck by the fact that since the advent of the Atomic Age the area "has always served a major function in luring nuclear physicists, super-computer designers, and aeronautics executives to its thousand-mile stretch. Isolation has, of course, been another great advantage, as laser beams, Stealth Fighters, and hardened tanks play out war games, and as malfunctioning missiles prepare to detonate with only a few ranchers around to complain."

They also point out that "the villain in this story is, and always has been, secrecy."

The same two authors comment that the nineteen-kiloton explosion of the Gadget "was not a terrifically efficient explosion: it didn't use up all of the plutonium in the core. So tiny bits of unexploded plutonium were spread over hundreds of miles."

An official 1978 study concluded that the area was "one of the significant plutonium contaminated areas in the United States, both in terms of the quantity of plutonium deposited and area extended."

Advocates of the nuclear option point out that Oppenheimer and his team could not have anticipated the extent of the damage to the population. The argument has some truth to it: the report on the explosion by Colonel Stafford I. Warren, chief of the Medical Section for the "Manhattan District," observes that "the test was several times greater than that expected by the scientific group."

Another estimate says it was four times greater.

Colonel Warren went on to report that "the cloud column mass and top reached a phenomenal height, variously estimated as 50,000 to 70,000 feet." He added that "the middle portion (of the radioactive cloud) moved to the west and northwest".

That brought it right over the community of San Antonio.

He speculates that the energy was in the range of ten kilotons of TNT. As we saw in Chapter One, Enrico Fermi came up with nearly twice that figure.

Reme Baca recalls that "descending fallout debris, described as sand-like dust, 'like snow' or 'like flour' covered the desert landscape. It coated fences and posts, building, roofs and clothes lines. It also rained the night after the Trinity blast.

"At the time in New Mexico history when many ranchers' and farmers' roofs diverted rainwater into cisterns and barrels (...) their water became contaminated with radioactive debris, including plutonium dust."

Nobody was told about that.

Contaminants in great quantity, including Iodine-131 (which affects the thyroid gland) and many other radioactive isotopes leaked into goat and

Chapter Nine - The Investigation

cow milk, game and chickens, and poultry eggs. Yet no protective measures had been recommended to the population, as could have been done under some pretext: Hadn't the noise and light of the blast been "explained" to the population as an accident at an ammunition depot?

It was only in 2008 (!) that a review by the Centers for Disease Control (CDC) confessed that internal doses received by Trinity victims were never part of a scientific and medical evaluation, even though Manhattan Project biological staff were on hand and must have been well aware of the dangers.

At least, by the time they dropped it on Japan, they had readjusted the pressure sensors to trigger the detonation at 2,000 feet altitude, to avoid causing the same long-term damage the bomb had caused when it exploded at home, in New Mexico. But no long-term epidemiological or genetic damage assessment study was ever initiated for the "downwinders" of Trinity.

In his remarkable book, *140 Days to Hiroshima*, written from the point of view of the Japanese, historian David Dean Barrett recalls that General Marshall had contemplated the possibility of using atomic bombs as tactical weapons, "to soften up the Japanese forces." He had reviewed the scientific and medical data coming out of Alamogordo and "he believed American forces would be better off facing the risks of radiation than the vast numbers of Japanese defenders."

Barrett goes on to note that neither Marshall, "nor anyone else in the US, including the Manhattan Project scientists, clearly understood what those risks entailed."

"In fact, *six years after the atomic attacks on Japan, the United States government was still using American soldiers and Marines essentially as human guinea pigs*, placing them a mere two miles from the epicenter of nuclear explosions and then minutes later, after the blast wave had passed, marching them to the detonation point..."

Never mind the six-year delay: Barrett's book highlights the fact that even seventeen years after Hiroshima and Nagasaki, the Pentagon was still ignoring the full dangers of radiation to health: "At the Nevada Test Site (...) American soldiers and Marines were placed in trenches only two miles away from the center of a nuclear explosion. After the blast wave passed over them, they walked to Ground Zero. The Atomic Veterans, as they became known, contracted, as a result of their radiation exposure, one or more of the twenty-one different cancers. Eventually, the US government paid out 800 million dollars in claims." (op.cit. p.278)

David Barrett concludes that what seems clear from the length of testing is "how little they apparently knew." We may discover that the same statement applies to what "they" know about UFOs, and especially about the few findings reported in the open literature about their material composition.

∧ ∧ ∧

The catalogue of unclassified physical cases where recovery of a material specimen was done under reliable conditions is small, yet it provides a basis for some simple comparisons. We have selected 11 of the most interesting studies, whose results have been published over the last 40 years: I had discussed these cases in detail as early as 1998. New cases happen and are described with uneven frequency, so our collection expands and we will continue to report on physical findings, so it is important to place the San Antonito crash in the perspective of these other known incidents.

Together, they form an interesting tapestry.

Case no.1: April 17, 1897 in Aurora (Texas)

The story, as told at the time in a local newspaper, stated that an unknown object "sailed over the public square and, when it reached the north part of town, collided with the tower of Judge Proctor's windmill (reports do have it as a windmill, but it was a windlass) and went to pieces with a terrific explosion, scattering debris over several acres." The ship had a pilot, the story went on, and he died in the crash. He "was not an inhabitant of this world," and undecipherable papers were found on his person when they buried him. As for the metallic object, it seemed to be made of "a combination of aluminum and silver."

My friend Don Hanlon and I published the story in 1967, treating it as interesting Americana. Our article reawakened interest in the incident: it was investigated again in 1973 by William Case, a journalist with the *Dallas Time-Herald*, and by personnel from the McDonnell Douglas aircraft company. Their site research uncovered surprisingly solid evidence: something very strange had actually happened in Aurora. (Note)

Note: McDonnell company report on the Aurora case, by Holliday, J. E., unpublished, dated August 13, 1973. The on-site investigators were Ronald A. and N. Joseph Gurney (May 12, 1973).

Chapter Nine - The Investigation

One fragment in particular was recovered from the well and another one, found 50 yards NW of the windlass, was found buried approximately 10 cm in the soil. The first fragment was in a molten state when it traveled through the air and impacted the ground. Another fragment from the well had also been in a molten state when it hit the wooden windlass.

What was unique about this well fragment was that the molten metal carbonized the wood when it hit the windlass structure and pieces of that carbonized wood, plus a square nail, became embedded in the metal.

The fragment retrieved by Mr. Case and his team was determined to consist of aluminum (83%) and zinc (about 16%) with possible traces of manganese and copper (sorry, no "silver"). That combination might originate with numerous common aluminum alloys, according to the McDonnell scientists, *but not prior to 1908.*

One long-term expert on the case wrote to me: "Due to the characteristics of these two metal fragments and where they were found, it would be hard to conceive that this was a hoax. Brawley Oates, a senior citizen living on the land by the well, handed us the fragments. He had no desire to get into the limelight, being more concerned about his health, since he'd been drinking the water from the well before it was capped off."

It hasn't escaped our notice that the Aurora object, like the oval craft seen by Padilla and Baca, hit a tower before it went crashing to the ground: two similar accidents, half a century apart... If those are piloted by sophisticated Aliens, what does that tell us about their navigation skills?

Case no.2: June 21, 1947 in Maury Island (Washington State)

On that particular afternoon (three days before the famous Kenneth Arnold sighting) four people who were on a boat close to the shore of that island located near Tacoma, reported an observation that has puzzled and divided researchers ever since. According to the published story, the witnesses were Mr. Harold Dahl, a salvage operator, with his 15-year-old son and two crewmen. They had a dog with them. They reported seeing a group of six large, flat doughnut-shaped objects flying at an altitude estimated as 2,000 feet. Their central holes were about 25 feet in diameter and they glistened with a gold-silvery color.

One object suddenly started wobbling and dropped to an altitude of 500 feet above the boat. Another disk dropped down (as if to help the one in difficulty, according to Dahl). A dull explosion was heard and numerous sheets of a light, thin metal issued from the central opening in the troubled object while the witnesses were showered with hot, dark fragments resembling lava rock or slag, compared to brass in color. The dog was reportedly hit by one of the fragments and died.

The shore was found littered with glassy material and silver foil.

Military authorities and the FBI, in a confused series of investigations, attributed the case to a hoax, a conclusion that has been embraced by many ufologists, but I don't think any such statement can be drawn from the reported facts.

As Mr. Ray Palmer commented: "There we have it. The samples (...) were not slag nor were they natural rock. What were they?"

Case no.3: July 1952 (Washington, DC)

Information obtained by well-known journalist Frank Edwards memorialized an incident when a metallic piece "fell" from a flying disk. A fragment representing about one third of its volume was examined a few years later by an official Canadian government researcher, Mr. Wilbert Smith.

Over one inch in size, it was remarkably hard and reportedly consisted of "a matrix of magnesium orthosilicate" composed of fragments measuring 15 microns.

Interviewed by two civilian researchers, Messrs. C. W. Fitch of Cleveland and George Popovitch of Akron (Ohio), Smith added that a US Navy pilot had been chasing a flying disk when "he saw a bright scintillating fragment detach itself and fall to the ground."

It was recovered an hour later and weighed 250 grams. Smith reportedly showed the sample to Admiral Knowles. We will have more to say about this fragment—and its remarkable origin-- in a later chapter.

Case no.4: December 14, 1954 in Campinas (Brazil)

Numerous witnesses are said to have observed three disk-shaped objects in flight over the city. Again, one of them started wobbling wildly and lost altitude. The other objects followed it down and it stabilized at about 300 feet.

At that point, the troubled disk emitted a thin stream of silvery liquid. The material was reported to splatter over a wide area including roofs, streets, sidewalks, even clothes left out to dry. An analysis by an unnamed

Brazilian government laboratory is said to have identified tin (Sn) as the main component of the samples.

An independent analysis by a private chemist, Dr. Risvaldo Maffei, reported that 10% of the material was composed of other elements, but gave no precise data.

Case no.5: November 11, 1956 in Väddö island (Sweden)

Professor Sturrock of Stanford University obtained a sample reportedly recovered by two witnesses of an aerial phenomenon. Although the material appears to be common tungsten-carbide, laboratory study by Mr. Sven Schalin at an SAAB aircraft laboratory, and also at unnamed laboratories in Sweden and Denmark, disclosed it also contained cobalt and was consistent with manufactured products.

Case no.6: Probably 1957 in Ubatuba, near Sao Paulo (Brazil)

This well-studied incident came to light through the efforts of a dedicated investigator, Dr. Olavo Fontes and of Jim and Coral Lorenzen of the Aerial Phenomena Research Organization (APRO) in Arizona.

The initial report stated that witnesses on the beach at Ubatuba reported seeing a disk that plunged towards the ocean at high speed, rose again to about 100 feet and exploded, showering the area with bright metallic fragments, some of which fell in shallow water or on the sand. A few fragments were analyzed in Brazil by Dr. Luisa Barbosa at a laboratory specialized in mineral production analyses. She identified the major components as highly pure magnesium, more pure than commercially-produced magnesium.

Subsequent work under the direction of professor Peter Sturrock has been conducted at Stanford University and at Orsay University in France, confirming that the material was magnesium and magnesium oxide, with a minute amount of "impurities," primarily aluminum, calcium and iron. Of course we don't know what role the "impurities" may be playing here.

Case no.7: July 13, 1967 in Maumee (Ohio)

A collision between a car and an unidentified light about 11:26 pm produced evidence in the form of two metal samples on the road and "fibrous material, also metallic, found on the car."

Both witnesses were Navy veterans, one of them a radar specialist. In their police report, they indicated they had unexpectedly encountered an intense source of light, compared to a welder's arc, in the middle of the road while traveling West on Stitt Road towards Whitehouse, Ohio. Swerving to the right, the car skidded for about 70 feet and the two men expected a collision, but no trace of the object was left when the car stopped. Some "fibrous" material was later found on the car, however. It was determined to contain 92% magnesium, according to Coral Lorenzen of APRO and the Condon report of the University of Colorado.

Case no.8: Early 1970s in Kiana (Alaska)

Professor Sturrock has obtained one of two recovered fragments of this unusual sample, found by an Eskimo on a river bank, following observation of an unusual aerial phenomenon.

Each specimen is silvery, light-weight, and looks as if it had been poured in a molten state from a source close to the ground. The exact composition is unknown.

Case no.9: 1975 or 1976 in Bogota (Columbia)

Two students from the University of Bogota were about to take a taxi about 4 am that night when they heard metallic sounds overhead. They reported seeing a disk, about 12 feet in diameter, swinging in the air as if it had difficulty maintaining its altitude of 3,000 to 3,500 feet, obviously a rough estimate given the conditions. Four objects appeared, flying around the first one as if to provide assistance. Spouts of liquid were then ejected from the primary object as the witnesses took shelter under a tree and watched the liquid fall on the pavement, producing a vapor. After letting the material cool down for ten minutes while the objects flew away, the witnesses were able to recover two metal chunks, about 4 inches by one inch, and a quarter-inch thick.

The first analysis was performed in Central America by an engineer from a petroleum company. He found the sample to be composed of an aluminum alloy with magnesium and tin, along with other, unidentified materials. The material was easy to cut, non-magnetic and presented very fine granulation.

In 1985 I obtained a sample of that material from Mr. Ricardo Vilchez, a Latin American researcher, and I brought it to the US for further study at the University of Texas, thanks to Dr. Harold Puthoff of the Institute for Advanced Studies in Austin. Indeed, we found it composed primarily of Aluminum (93.7%) with phosphorus (4.8%) and iron (0.9%) with traces of sulphur and an unexplained oxy-carbide layer. Subsequent SIMS (scanning ion mass spectroscope) analysis found aluminum and magnesium along with potassium, sulphur, sodium and silicon. Phosphorus and iron showed up as trace elements (no tin).

Case no.10: December 17, 1977 in Council Bluffs (Iowa)

In this incident, which my colleagues and I have analyzed in detail, two residents of the above town, an eastern suburb of Omaha (Nebraska) first saw an object fall to the ground at 7:45 pm in the vicinity of a dike in Big Lake Park, on the northern city limits. A bright flash was observed, followed by flames 8 to 10 feet high. When the witnesses reached the scene, they found a large area of the dike covered with a mass of molten metal that glowed red-orange.

The incident is of particular interest because police and firefighters reached the scene within minutes and were able to identify and interrogate eleven witnesses in all. Two of the witnesses had independently seen a hovering, red object with lights blinking in sequence around the periphery, hovering at low altitude. The recovered sample was analyzed at Iowa State University and at the Griffin Pipe Products company, leading to the determination that the metal was chiefly iron with small amounts of alloying elements such as nickel and chromium, excluding a meteoritic origin or indeed any other natural occurrence. The incident has never been explained.

Case no.11: circa 1978 in Jopala, near Puebla (Mexico)

The fall and recovery of a metallic residue from an unknown flying object was reported to me by local investigators while I did research in Mexico in November 1978. The object, which fell in the mountains near Puebla, was found to be composed of iron with silicon (1.13%) and traces of manganese (0.84%), Chromium (0.77%) and carbon (0.28%).

Table 2: Summary of Sample Composition

	Case		"Slag"		"Light silvery alloy"	
No.	Location	Year	Primary	Secondary	Primary	Secondary
1	Aurora	1897			Al, Zn	Mn, Cu
2	Maury Is.	1947	Ca, Fe, Zn, Ti	Si, Cu, Ni, Pb	Ag, Sn, Cd	
3	Washington DC	1952			Mg	
4	Campinas	1954			Sn? (90%)	
5	Väddö	1956	W(94.9%)	Cu, Zr, Fe		
6	Ubatuba	1957			Mg	Al, Ca, Fe
7	Maumee	1967			Mg (92%)	
8	Kiana	1972?			light material	
9	Bogota	1975?			Al(94%)	P, Fe
10	Council Bluffs	1977	Fe	Ni, Cr, Mn, Si, Ti		
11	Jopala	1978	Fe	Si, Mn, Cr, C		

Note: Since this study was published, our team has re-investigated the nature of the material recovered in case 6 (Ubatuba) and we have studied the isotopic ratios in case 10 (Council Bluffs).

See Nolan, Vallee, Jiang, Lemke: "Improved instrumental techniques, including isotopic analysis, applicable to the characterization of unusual materials with potential relevance to aerospace forensics" (*Progress in Aerospace Sciences* 128 [2022] 100788).

Other incidents have been reported in recent years, with more physical samples from South America, Russia and France, among other countries.

Why was that extensive body of precise data ignored so long, while metallurgy is moving on with new generations of "high-entropy" alloys with remarkable properties?

And why do scientists, even today, insist to remain studiously ignorant of its existence?

In addition to analyses of the chemical composition, some advanced work has been done on isotopic ratios within the elements found in these samples. In particular, some authors have suggested that the Ubatuba magnesium had been artificially altered. My colleagues and I are re-doing these measurements with more precise methods to find out the truth. We have applied the same techniques to the Council Bluffs samples, and we expect to publish those results in the scientific press, after proper peer review, in an effort to place the subject within the "official" scientific literature, where it has never been admitted. But we're a long way from the caliber of investigation the subject deserves.

CHAPTER TEN

THE DEVICE AND THE DREAMS

After a day on-site, and a review of the new data from Mr. Padilla, Paola and I met early the next morning and we drove over to an out-of-the way coffee shop to study "The List" and plan the day. The next batch of questions on that list had to do with the exact positioning of the "device," and the way it was retrieved by the child. I had become convinced it was likely a common industrial object, but Paola disagreed, and indeed some mystery remained. Actually, it had only become more puzzling: How did that piece of rough aluminum alloy end up pinned to the interior wall of the extraordinary vehicle the military had been so eager to remove and hide away in an undisclosed laboratory, where it presumably rests today?

Armed with a new set of data, we went back to Jose to try to answer that question.

Paola: OK, so did you feel like you had to hurry up? And pull that thing off the wall?
Jose: No…See, the object was kind of leaning, on an incline, where that mechanism there, was on the wall, and as I was standing, I could be

looking at it on a slant. That's why I put that cheater bar and the pin fell on the floor. It was almost not upright. (Note)

Paola: So the pin fell to the floor, because it was leaning over... it was not upright. What I asked was, were you nervous because a sentinel would find you? Weren't you in a hurry?

Jose: No.

Paola (*laughing*): Why? Weren't you scared?

Jose: (*very serious*) No. I've never been scared.

Paola: So, you felt you could take all the time in the world?

Jose: Well, whatever was going to happen, I was there, and I was going to do my thing.

Paola (*laughing*): OK, OK Jose, but you were never worried you were going to be discovered?

Jose: No: everybody was gone! There were no Aliens there, the soldiers were gone from there, so I figured we had all the time, to do whatever we were going to do.

Now we had a fairly complete picture of the events that led to Jose and Reme walking off with an artifact from inside an object that, for lack of a more scientific term, has to be called a UFO.

No other witnesses, in recorded history, have actually done that before them, or since. Any significant material associated with these objects is in the hands of governments, and off-limits to researchers without a specific clearance, *and* related need-to-know.

As far as Mr. Padilla's comment that he was "never scared," that statement may strike some readers as boasting or an old man's self-indulgence. Over multiple meetings with him, I have built up a different opinion: His calm and patient attitude with us was deceptive, because there was so much underneath that was built on unusual strength, courage and a knowledge of people good and bad that only an experienced police officer can acquire.

This became apparent when Paola and I became very concerned about an imminent surgical operation he told us he might have to undergo, a procedure to retrieve a bullet that had lodged itself in his abdomen as he was arresting a criminal in California during his active years. At his

---

Note: What Mr. Padilla called a "cheater bar" is what is now commonly called a "crowbar."

Chapter Ten - The Device and the Dreams

age, anesthesia itself could be problematic. Further review of the bullet's placement with respect to vital organs, fortunately, led the surgeons to the realization that everything could be left alone, to everybody's relief. But the patient himself showed no sign of anguish.

^ ^ ^

Next, we discussed the matter of Jose's recollection about the beings themselves. Would he agree to undergo professional hypnosis (not the amateurish kind) if it might enhance his recall of that first evening?

Jose: Hmmm, no, my memory, you know… I remember very well what I seen and the thing is that, …I wish I would have gone in there (*note: on the first day*). Maybe I could tell you more about it. But I didn't go inside there and meet these, these kids, or whatever, you know. That is all I've seen; and what I've seen, you know, I still remember every bit of that.

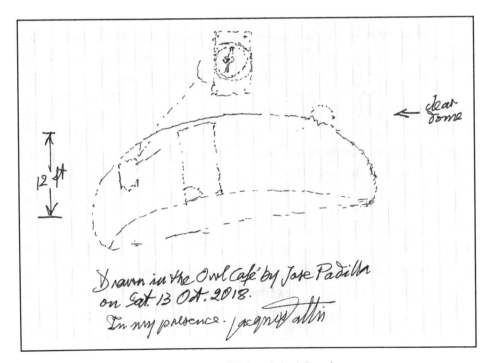

Fig. 21: Mr. Padilla's original drawings.

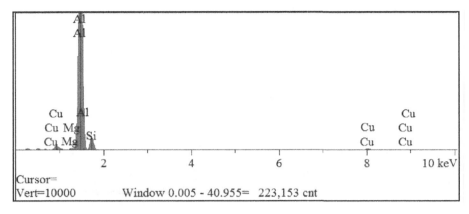

Fig. 22: The simple chemical composition of the metal bracket.

Paola: Yeah, I know. Jacques is struggling with the fact that you said it looked like a "fire ant" and then you say it looked like "little men." So, the fire ant, is that just the description of the long arms, or were they insectoid people?

Jose: The thing is that ...what I meant by them looking, "like a fire ant," it's, a fire ant farm which I had. That I bought for the kids one time. And the way that they stand...I stood right there and watched them work. They work 72 hours a day, you know! (*laughing*) And the way they work sometimes, they stand up, in an upright position and that reminds me of something, you know, like these people were.

Paola: But they, but ...you said "these people," so you think they're people, instead of ants?

Jose: Yeah, they were people. But I picture them, that the way they moved, was like a fire ant, OK?

Paola: OK. Alright. Now, one thing that I'm curious about, Reme told me that he saw a vision of people falling out of skyscrapers, that he had a vision, they flashed something in his brain. Do you remember that?

Jose: No. I...used to dream. But lately I haven't, I haven't dreamt. But I used to dream after that, that, uh, there was, like, uh, clouds, falling from the sky. And then I used to wake up perspiring, and that used to erase the whole thing. A few times that I dreamt about, there was, was some...like clouds or something, that were falling from the sky, but they never touched the ground. I used to wake up and everything used to, uh, disappear.

Chapter Ten - The Device and the Dreams

Paola: Was that scary?

Jose: No.

Paola: Why were you perspiring?

Jose: I believe it was, uh, a dream, you know, of what I seen, that had happened. And at that time, it seems to me like, a dream like that, is like when someone is calling and you can't get to them to help them.

Paola: You said "clouds were falling from the sky?"

Jose: Yeah, clouds were falling from the sky. But I was waiting for them to disintegrate or something like that, but then they hit the ground, and I used to wake up perspiring.

Paola: What do you think those clouds could be?

Jose: I, uh, I don't know...since that...

Paola: Since that...go ahead?...

Jose: It could have been that, uh, that something, that these people were putting, um, so like a dream...nothing happened.

Paola: I don't want to interpret your dream or anything, but the clouds, when I think of clouds in your area, I think of mushroom clouds or atomic experiments which happened in Nevada after that. Or did you think they were regular clouds, or could they have been testing clouds?

Jose: No these were clouds that...like what the lightning was striking in there, and it was raining, and we were, uh, taking shelter underneath that cliff. That's the way it seems to me like, the clouds like that, they were kind of breaking, but they never hit the ground.

Paola: So they were lightning clouds, in your dream--was there lightning?

Jose: It was lightning, lightning clouds, but it...it seems like, uh, I must have been dreaming pretty hard or something, I used to wake up perspiring there. Something woke me up!

Paola: OK, you never saw the creatures in your dreams?

Jose: Uh, no.

Paola: And how old were you when you had these dreams?

Jose: I was...nine, ten, eleven years old.

Paola: And, Jose, did you ever see, when you were living there, did you ever see anything in the sky, like a UFO, or have a sighting? Or look up in the sky and look for them?

Jose: Uh, no. We never... I never thought about anything, what just happened, you know. This was something that happened, that I experienced, that I'd never seen before.

Paola: So you didn't, like, look up and see if they were coming back? Or, look up in the sky, to... in other words, it was a one-time experience for you, and then you just let it go?

Jose: Yeah, I did. I did. But sometimes I used to stop and think for a while, that, what happened there, and then, our... *It was a day and a half just when we came back, which was too long, that we stayed (away) too long.* I was thinking, you know, that, maybe they had somebody there, that came over and picked them up, these creatures.

Paola: Oh, well, I've heard stories like that too, I've heard stories, you know, that their own come and get them. Yeah, I've heard that too.

Jose: Like scouts... they have scouts that they, uh, come and pick the object or little men, *or whatever.* They come and pick them up, and that's probably what I thought that after we went back over there and, uh, it was cleared up, like they had cleaned it up with a rake. To me, you know, the first thing in my mind, about that....someone had come over and picked them up!

Paola: Did they throw brush? They didn't throw dirt on the craft? They threw brush, right?

Jose: They threw uh, pieces of brush – they were broken.

Paola: And you said that the mesquite trees were much bigger, so they threw pieces of mesquite.

Jose: Yeah, the mesquites there, were almost trees. The cactuses there were taller then, taller than us! And as you've seen, you've been there a few times, you have seen how the cactus are...they are real small. The mesquites are really all dead. (Note)

Paola: So you think that the changes in a lot of the flora there, are because of the crash?

Jose: Yeah, that's right.

Paola: Do you believe that? Or, I mean, is that your thinking?

Jose: It was my thinking.

Paola: That area is different because of the crash, and I know when I went there with you, I could see the circle perfectly, and nothing grew in the middle.

Note: This change in the landscape, which we also noted on our way to Ground Zero, is a lasting effect of radiation, not of the UFO crash. It is important to keep in mind that the San Antonio area essentially bears the wounds of a war zone, devastated by a full-scale nuclear explosion, not just a "test."

Chapter Ten - The Device and the Dreams

Jose: Yeah, well, I got the picture of you and me at the circle...it's still right there. I had a picture of another thing that my son took off from there, while I was walking around there and that circle was there, and then all of a sudden these, green plants started coming out, you know. All these...things that have happened, as long as I have lived there, uh, I have never seen anything like that before.

Paola: And we're talking seventy years, almost...

^ ^ ^

As I listened to Jose Padilla, trying to reconstruct the situation he faced and how it affected his life, modern Silicon Valley seemed very far away in my mind, hidden behind a fog of speculation, probabilities and potentials, all based on interesting hypotheses about reality. Jose's reality, in contrast, was immediate and terribly obvious: Something was wrong with the well-behaved way Academia had been teaching us about the world, and looking at the things that didn't fit accepted models.

Back in Palo Alto or Paris, if you are a scientist, an engineer, or any kind of technical worker, people at jolly cocktail parties are likely to approach you with questions about UFOs.

You are supposed to put down your drink, wipe any lingering smile from your face, and seriously recite from the accepted textbook: Yes there are sightings, and those cops and pilots are truly good people, but they can be fooled, you know, like all of us, by all those optical phenomena in the sky, not to mention fancy secret prototypes from the Pentagon. It will all be revealed in due time...That's the answer people expect, because they want reassurance.

And none of that makes any sense, when you look at the real data.

In a recent promotional review of an amused but skeptical book about UFOs by Sarah Scoles teasingly entitled "They are already here," a major science magazine could only agree with its author (Note): "It makes for a good story with worthwhile insights about humans, but sadly no real space aliens."

Never mind that the well-worn "space alien" idea from last century Hollywood is only one of many hypotheses about UFOs, or that the same magazine's other pages are full of insights and well-articulated theories

---

Note: *New Scientist*, December 5, 2020, page 30.

every week, about potential forms of life and consciousness in the universe that could manifest in infinities of ways, all of which admissible under today's official science. In that context superficial dismissals of UFOs, and the reviews that embrace them in another layer of puffery, can be pointed at as the real reason for the long silence of many witnesses, like the ones Paola and I have met in our own research over the last few decades.

There, the missing data lies.

The magazine didn't even flag the blatant mistakes in the book they were reviewing: they lazily repeated the false data that "the first flying saucers were reported in 1947 and became part of popular culture despite the lack of scientific evidence for their existence." In reality, witnesses of early cases like Mr. Padilla are still alive and can testify about the physical objects they observed and the material recovered in 1945 and even before. They did it under circumstances that never made it to the mainstream of what academics like to call "popular culture" (Popular... as opposed to what? Doesn't all culture come from people?)

A simple review of the research literature on UFOs—yes, there is such a thing—would have disclosed careful studies of that question. To cite just one example, a paper presented at an AIAA (American Institute of Aeronautics and Astronautics) conference held in 1975 found 52 witnesses of close encounter cases *before 1947*. The number of known "early" witnesses has grown considerably since that survey was published, proving that we are not dealing with a spurious sociological effect of modern technical dreams (Poher & Vallée, 1975).

For those who take the trouble to look at the real reports, vetted by serious field investigation and brought inside the lab, the phenomenon reveals itself to be infinitely more complex--and more interesting.

∧ ∧ ∧

Now I was back in San Francisco, having changed into city clothes after washing off the dust of New Mexico and the smell of mesquite. In passing, I had learned that Native Americans used the wood from those mesquite trees for incense, believing the scent would attract good spirits.

What we need most urgently in this research, I thought, isn't a lot of money, or dumps of every uncorrelated digital signal from space radar, or Navy infrared cameras, but serious intellectual investment from the scientific community. Those are the "good spirits" we want, but they have

Chapter Ten - The Device and the Dreams

ignored the problem and they don't know anything about the available data. So I decided the most useful contribution I could make was to help build bridges.

The opportunity came when friends from Silicon Valley invited me to a private evening reception, to be followed by a lecture by a well-known *New York Times* science journalist. He was about to interview a Stanford professor about the latest developments in Artificial Intelligence.

There were over a hundred guests, distinguished donors and members of a reputed non-profit organization that sponsored deserving young students throughout the country. The crowd was smart and generous. It was also elegant and festive: Silicon Valley recent wealth mixed with Old Money from deep California roots. As cocktails were served ahead of the lecture, I was introduced to someone I had hoped to meet for a long-time, an award-winning academic with a national reputation as a leading educator and proponent of science.

"This is my friend Jacques, long-term colleague of mine," said my guide, pushing me forward; "his background is in AI and astrophysics" (the professor granted me an indulgent look, a hand was thrust forward)... "but he's best-known for his research about unidentified flying objects..."

The professor stopped in mid-smile; the hand was retracted; blood went to his forehead. His eyes looked away: *What if he was seen in my company?* He had frozen under the shock of those words. I thought he would throw up, but he regained his balance and blurted out: "Ah yes! Like that teenage girl I saw on TV, abducted somewhere in the Midwest, captured by ETs. Now she's pregnant...she blames it on the Aliens!"

A few people heard his eructation, turned around and giggled.

"And that's the kind of sick garbage people believe, these days!" he concluded as he walked away.

I looked around in confusion.

"So that's all they know these days, in the Academy?" I asked the financier who had introduced me, as I looked for the exit from the depth of my humiliation.

A couple of well-coiffed ladies, passing by, stopped eating their zakuskis and stared at us.

CHAPTER ELEVEN

BACK IN THE LAB

The Padilla case is unique in the annals of this research for no less than *ten* reasons:

(a) Two witnesses *were on site while the craft crashed.*

(b) They knew the area intimately.

(c) They had covert access to the scene every day except Day Two, until the craft was removed.

(d) They were able to approach it stealthily before it was hauled away.

(e) Three of the witnesses, including Jose Padilla, his father, and a State policeman, actually spent time inside on different occasions.

(f) An unusual metal device was detached and recovered from the interior wall of the craft.

(g) We had that device available for our own, detailed inspection at the time of the writing of this book.

(h) One of the two witnesses, Mr. Padilla himself, is still alive with an intact memory at age eighty-three after a long career as a California State Trooper.

This gives the case heightened scientific interest, even compared to Roswell or other crash cases in the literature. In addition, please consider that:

(i) Paola Harris met Reme Baca and his family at length ten years ago, recording a very detailed interview with him at his home.

(j) We have had convenient access to the site multiple times during this research, spending several days each time with a search team and instrumentation. (Note)

In a footnote in his book, *Majic Eyes Only*, Ryan Wood observes: "It is not clear how close Jose and Reme were to the crash site or military perimeter. Either they were very lucky not to be discovered by the military perimeter guards, or they were very far away."

It was reasonable for Ryan Wood to make this assumption at the time, but from our own records, none of that was the case. On the contrary, as we have seen throughout this investigation, the two kids (who were on site fully *two days before the soldiers showed up*) took advantage of their detailed knowledge of the land and their desert experience to remain hidden. One could not have hoped for better observers. They were equipped with binoculars which they knew how to use every day with cattle, and they escaped detection in the rough terrain. The kids also noted the patterns of the military movements: they could guess how long the young soldiers would be in town for food and drinks, occasionally, as on the very last day, completely relaxing their surveillance since no credible threat existed to their mission.

Above all, the kids made outstanding observers, even at that early age. And no wonder: Jose's father expected them to regularly take a count of the

---

Note: Three knowledgeable investigators: Timothy Good, James C. Fox and Mexican journalist Jaime Maussan, also recorded interviews with Reme Baca, confirming Mrs. Harris' information.

Chapter Eleven - Back in the Lab

cattle and even ascertain the brands at great distance. They moved easily on horseback over terrain that even a Jeep couldn't navigate. They were used to tracking wildlife in the area, a very large property of thousands of acres.

Thanks to our multiple trips to the site, we also know exactly where they stood when they spied on the Army detachment. In our follow-up with Jose, we established the fact that they entered the area from the back (rather than from the gate) and crawled up the ravine that can be seen on figure 18. By hiding there, among the bushes and small trees, they could watch every movement, often from as close as some 200 feet without being detected.

Jose also recalled that they sometimes came from the West, and by lying flat among the mesquite behind the top line of the hills they would have a plunging view on the recovery scene. If the soldiers saw them (they probably did at least once, since there is a reference to "a couple of Indian children" in the military records, reported by Lt. Colonel Brophy from his aircraft), they never imagined that the kids would comprehend the significance of what they were doing.

If the Army was simple-minded enough to try to fool the parents with some silly tale about weather balloons (why would you need to bring in a large tractor-trailer and a crane to load a weather balloon you could just have deflated?) they must have thought the two "Indian" urchins could never have any concept of what the disabled craft was.

Their own commanding officers were never aware that the kids were eagerly watching their movement with field binoculars every bit as good as the military version.

∧ ∧ ∧

Three analyses of metal pieces associated with the crash have been reported before ours.

In his November 2005 book *Magic Eyes Only*, which mentions about 93 claims of UFO crashes, Ryan Wood dedicates a six-page chapter to Reme Baca's recollection of the 1945 San Antonio case. He does not disclose when he spoke to him, and we do know the two never met, but he mentions examining "a piece of apparent crash debris retrieved from the site by Reme Baca and Jose Padilla" (Note *next page*) that Reme was reluctant to discuss in detail.

Reme stated he had submitted it to various science organizations for further study: "In addition to local tests conducted in Washington State, we began working with Dr. Smith, of the Metallurgy department at an international facility, where three different electron scanning microscopes were used to look at the metals and analyze their elemental makeup. The first thing we found is that the metal appears to be essentially an aluminum silicate."

There was more to the story, however, because "Dr. Smith" apparently found an unusually high percentage of carbon in the sample. Reme Baca reports cutting off some pieces and polishing others to take a better look, which led to the discovery of what he calls "weird structures":

"We got some great pictures of very weird structures in the metals. They looked like little skeletons of bugs squashed into the metal. Also, there are short strands of some other material that we believe may be carbon fibers. This sort of 'hairiness' about the metal provoked a lot of questions from the academics and electron microscope operators."

The description goes on with even more sensational claims: "A blend of carbon and *some other trace materials (?)* is used, which dramatically increases the conducting power, eliminating the resistance to electricity while at the same time a transference of heat takes place. There appears to be the potential for heat shielding, or computer chip manufacturing."

Since we can't be certain who did the tests or where, and none of the "great pictures" have been published, the claimed discoveries by "Dr. Smith, plus a geneticist and a scientist with a metallurgical background" remain at best intriguing data, at worse anecdotal boasting.

Reme Baca died in 2013, so we didn't have the opportunity to follow-up on the details of that analysis. Fortunately, we do have the well-supported conclusions of dedicated investigator Timothy Good who deals with the topic of analysis in his very detailed book *Need to Know* (**13**). He recalls examining and photographing the boys' *souvenir* (i.e., the famous 'Tesoro') when he met with Mr. Baca in Seattle on June 6th and 7th, 2004:

"It appears decidedly terrestrial," he wrote, "a point conceded by the witnesses. Looking like a bracket of some kind, it is 12 inches long, weighs 15 ounces and contains a number of holes for fasteners of some kind. A section cut off for analysis, as well as acid tests, reveal the metal to be

---

Note: Reme Baca wasn't telling Ryan Wood the whole story: Jose Padilla is the one who bravely climbed on the trailer where the craft was waiting for shipment. He is also the one who found the metal piece near the ranch gate, and gave it to Reme Baca for safekeeping.

200-series aluminium, or similar. A smaller, semi-circular piece, believed by the witnesses to have also come from the craft, was found years later at the ranch entrance. Tests indicate 330/380 series aluminium (or similar)."

A third study, an extensive Australian investigation done in 2016, **(27)** also found that the metal sample was not radioactive and that it had two main components, confirming Reme's statements. Mr. James Rigney and his colleagues spoke of it containing "Aluminium and Silicon. Both are in hyper-eutectic alloys 4000 series form, also known as 'Silumin' commonly used in industry, auto, military and aeronautics."

They concluded:

"The analysis shows the sample to be terrestrial."

Their test, shown on Fig. 22, confirmed the object was composed of 87.06% aluminum, 10.45% silicon, 1.97% copper and 0.53% magnesium, typical of a dye cast aluminum alloy AA 383.0 (therefore, probably not in the 200 series as reported to Timothy Good).

Silicon is commonly added to Aluminum to lower the melting point of the metal without making it brittle. The report adds that the sample had likely been sand-molded. That means it could have been produced on-site by an Army technical group for its own needs, which would explain why it bore no brand or standard industrial markings.

A follow-on study found a composition of 87.09% aluminum, but only 2.70% silicon, 1.83% copper, but also carbon at 2.03%, iron at 0.94%, zinc at 1.66%, calcium at 1.19% and manganese at 0.54%. Magnesium is not part of the laboratory findings.

Yet another piece, which is in our custody, was cut from the bracket at Paola Harris' request by Mr. Sid Goldberg, a Canadian researcher, working along with a craftsman who stated it was "too soft to be aluminum," and it didn't smell like aluminum when cut.

At the end of the day, the various analyses generally do agree the material is an ordinary Aluminum-Silicon alloy, with common traces of other elements. "The sample is observed to be standard and unremarkable," concludes the Australian study with no ambiguity, putting an end to the tales of extra-terrestrial high-tech, yet leaving us with lots of subsidiary questions.

The shape and possible function are generally reminiscent of devices activating mobile vanes, such as the mechanism of water wells, and the mast section of windmills, for example in the Aermotor Company's list of standard parts widely used in industry **(28)**. Yet they don't correspond to any of the specific products we've found so far. By the same token, it might have come from a military device or an aircraft. As for the "smaller piece" mentioned by Good, it was a fragment of an I-beam that had some writing on it. That object was three feet long. It was found in 1953, as we have seen, by a cousin of Jose Padilla, who disposed of it as scrap metal, selling it as iron, an unceremonious end for an object once thought to have originated in the far end of the Cosmos.

∧ ∧ ∧

Saying that the "souvenir" is an ordinary human object does not answer our basic questions. None of the analyses we have reconstructed draws the final curtain on what Paola and I have jokingly started to call *"The Mystery of the Silumin Bracket."*

That convoluted instrument may appear standard but it is not "unremarkable." We have yet to ask what an ordinary, human fragment of some low-tech aluminum gadget was doing aboard a fantastic craft dropping from the sky in the middle of a storm, shattering the Marconi Tower of the White Sands Missile Range as its crew of diminutive insectoids skidded weirdly through the cabin.

Recall that the bracket, which is shown "as is" on figure 20, was not connected to any rods, hooks or pullies, as its prominent features and holes would seem to indicate for its primary function. By the time Jose "liberated it," it was simply spinning in a copper circle, held to the wall of the craft by one fairly loose pin that a nine-year-old kid could pop out. Could it be that the device discovered by Jose Padilla was not part of the vehicle's original equipment, but more prosaically an improvised spool hastily assembled by the soldiers using items they found on-site? Perhaps they needed something as simple as a device to wrap up an electrical cord? Perhaps it was a shattered part from the destroyed radio tower, conveniently picked up nearby and repurposed by the Army as a winding tool?

Other items described by the boys suggest human (not "Alien") construction: Around the panel that was torn apart by the crash (a surprisingly minor element of damage, by the way, for such a long and tumultuous landing), they saw "odd pieces dangling everywhere, pieces that looked like the angel hair that we used to decorate our Christmas trees."

Chapter Eleven - Back in the Lab

Some researchers would later speculate about optical fibers, but the description reminds us more simply of common strands of asbestos or insulation.

At the time of the recovery, the kids saw the soldiers erect an L-shaped (or A-shaped) frame: "They tilted it to get it to fit the wreckage over the trailer, because it bulged out over on one side. The ripped portion was to the bottom outside of the frame." More details became visible: "This was a smooth surfaced object, the bottom half was darker than the top," wrote Reme.

The Marconi tower, which appears to have dominated the northern section of the White Sands Missile Range, reportedly generated its own power with a small windmill, so perhaps the specially-designed aluminum bracket came from there? That would explain why the device was still inside the craft on the last day, unlike other materials the Army may have found there.

This is only one of several hypotheses we entertain, but it would also account for the fact that the military desperately wanted to get it back, pointedly returning to question Faustino about the whereabouts of their equipment, later sending a team to look under the floorboards of a house and even coming back with vehicles a few weeks later to search for... something the Army had misplaced.

That's a lot of military activity for one ordinary aluminum bracket. Recall that some diminutive creatures, independently described by the terrified Mexican sheepherder in similar terms to those used by the kids, transected through the wall of his cabin to demand the "Tesoro." It seems that everybody in the solar system suddenly wanted to recover that common piece of aluminum alloy.

I have that famous bracket in front of me as I write this. By the standards of Silicon Valley it isn't worth much more than a hubcap from your neighbor's used truck. In physical space, it is a common object. So why does it seem to acquire such importance?

The bracket may have been uninteresting from a scientific point of view, but it could have exposed the true scope of the recovery work if too many people talked about it and exhibited it. It might be evidence of military involvement in an operation the Army wanted to close once and for all, as the uneventful collecting of an ordinary weather balloon: they would successfully use that same explanation two years later at Roswell.

But you can't nail an aluminum bracket to the interior wall of a weather balloon, no matter how sophisticated.

Every nine-year-old kid in New Mexico knows that.

So what was it doing there?

∧ ∧ ∧

Now imagine a room, or a laboratory, somewhere in Silicon Valley, where the various pieces of the puzzle have been re-assembled by our small team: The military charts, the highway maps, the pictures from the time. You would also have the bracket itself, various 3D models of it, the microphotographs of the metal and the records of the three principal metal analyses, pinned to the wall. There would be a preliminary reconstruction of the energy calculations, the likely path of the object and the physical damage on its way to the crash. (Remember the extreme heat that deterred the kids as the vehicle plowed a long, wide trail in the ground). And then, in parallel columns, by date and time, all the statements made by those who went there or interviewed the witnesses before us, listing who did what, which vehicles were involved, and what questions were left unanswered by the records Paola and I were able to assemble.

In that room you might see another large display, showing the terrain where the object rested, and the changes that affected it as the ground level was raised over the years. Local traditions speak of it as a place where Indian tribes gathered because of the convenient proximity of the creek: there was an Apache ritual site there, a sacred circle of boulders, next to a hollow that may have been periodically flooded, providing water for horses and cattle. The area was gradually filled in with dirt by local authorities to control water flow. You certainly couldn't water any horses there when I first visited the site.

Ben Moffett's article in the *Mountain Mail* (**20**) also referred to the crevice where some of the material from the craft (Moffett calls it "odds and ends") was thrown in, which was "recently covered" (hence, circa 2003) by a bulldozer during flood control work.

It's still an active area: Between my two visits to the site with Paola, as we've seen earlier, someone had again sent a grader to rebuild a dirt dam or berm, in order to stop mud slides that could affect the whole area downstream and even cutoff the Pan-American Highway. In the process, the grader unceremoniously razed off the crash site vegetation, conveniently

Chapter Eleven - Back in the Lab

taking out half of the poisonous vegetation somebody had planted there before...all of which unfortunately made it nearly impossible for us to recover any small pieces of metal that might be buried there, now 20 feet under the surface.

Also on our "Analytical Wall" you would see a history of the Army's intervention, by days and personnel. Where did the detachment come from, and where did they take the strange object? This was the first crashed UFO recovery in post-war history, so there was no pre-organized unit and no standard protocol.

So the Army improvised.

First, according to Reme, they sent vehicles that could navigate the rough terrain. Mr. Padilla only recalls one Jeep, not two. At that point, from the point of view of a military officer, based on the testimony of two excited little kids, perhaps it was indeed "just one of our experimental weather balloons." But there was always the legitimate concern with atomic espionage: Somebody had to check out every incident.

Let us make a note here: We already know, at this point, that the object was not a craft from the United States Inventory, having encountered an accident during a storm and crashed on somebody's land. Nor was it a German or Japanese device with the kind of equipment that would have instantly escalated the discovery to an altogether different level. Nobody knew what it was, or cared, initially at least.

In this case, as in the later incident at Roswell, local authorities had *no details about what had happened* until Faustino Padilla, Jose's father, called them about the crash and took them to the site. Alerted by the pilot, the Army had his report but they were not missing any experimental device. They had noticed the damage suffered by the Marconi tower but the powerful thunderstorm that day might explain it. They only had the initial report by Brophy about an unusual object amidst the smoke.

In an area of the New Mexico desert that was monitored day and night by three late-model radars to detect potential enemy reconnaissance around the atomic site, the very fact that an anonymous intruder could fly in with an oval craft over 25 feet long in the middle of a storm without being detected raises some interesting questions. It seems to imply either electromagnetic stealth (a concept that was still far in the future) or a very short flight: *Where did it come from?*

∧ ∧ ∧

Let's move the calendar back to Saturday, August 18, 1945. Now the Army is well aware of the incident, both through Faustino's alert and also because of the damage to the Marconi tower, although they may have thought, initially, that lightning had struck it.

The Army scouts aboard the Jeep(s) would have made a report at that point, escalating the urgency of the case, but they would have gotten there too late to see any sign of the *hombrecitos*. Perhaps they were the ones who covered the object with some shrubbery, as best they could, to hide it from anyone riding by, until they could receive precise orders. But here again, there was no immediate intervention with a pre-cleared protocol, no arrival at the crash site of specialists in white hazmat equipment, none of the scientific reconnaissance with precise measurements, photographic documentation and environmental precautions we would see today, with Intelligence personnel recording every detail and doing rumor control.

Mr. Padilla recalls seeing an airplane circling around, confirming what we know from an Army Air Force pilot of long experience, namely Brophy (**29**) who left important details.

The recovery seems to have been left to a local unit that happened to be available. They began improvising at every step. From where we stand, they did a pretty commendable job.

Based on their report on the size and the weight estimate of the object, the military evidently decided it was important to retrieve the craft, secure it and examine it, which meant sending a large truck and a vehicle with an A-frame. For that, they needed permission from Mr. Faustino Padilla to tear off and replace his narrow gate and to send a grader to build a decent road, another couple of days of work. They also made it very obvious he had to keep his mouth shut.

In other words, as we reconstructed the sequence of events in Chapter Two, the crash most likely happened on Thursday, the 16th before the eyes of the two boys, who reported it that evening. The next day, Friday, they were kept away, helping Faustino (who had lost no time in calling authorities) with gardening chores, and the kids only drove back to the site on Saturday the 18th in the family truck, guiding him and State Policeman Eddie Apodaca to the actual spot. At that time, Day number Three, they watch as the two men climb into the object, which by then is empty of any creatures, but somebody has been doing real cleanup and raked the ground around the crash site. Presumably, Apodaca promptly reported his findings.

Chapter Eleven - Back in the Lab

If the Army came over on the 19th to see Faustino and to obtain permission to cut the gate, it wouldn't be until Day Five (Monday, August 20th) that they could move in with heavy equipment, although the Jeep might have returned to the site the day before then, to begin the detail work.

As far as the Army's mission went, the situation was as normal as one could manage, when you were sleeping next to an avocado-shaped device that had somehow dived down from the sky. By Friday, August 24th, the craft was covered with a light-blue tarp and secured to the tractor-trailer, ready to move the next day. A likely date for the craft's arrival back at White Sands is Saturday night, August 25th.

^ ^ ^

There is another fantastic element of the case we must take into consideration as we try to objectively summarize the known facts. The Mexican sharecropper hired by Faustino to deal with the sheep, Sr. Pedro Anaya, a lifelong friend of Alejandro Baca (Reme's father) had no interest in technology and little awareness of what had been happening in San Antonito. So his independent testimony, as a terrified witness of a related incident, makes it an especially important component of the story.

Recall that after "liberating" the bracket from its place on the wall of the crashed object, Jose kept it for a few days, then he heard that the Army was asking his father for permission to look through the house and tool shed, so he slipped the stolen item to Reme, who hid it under the floorboards of his family's storage room across the street. In Chapter Four, we mentioned that the military did go through the Padilla home, and walked away with some metal items, registration material that Faustino had stored there, and even some old weather balloons that had drifted onto the property.

Since ranchers in the area were used to picking up the soldiers' trash in the form of errant weather balloons that the military used in atmospheric soundings, and periodically returned them, I must say I find this "explanation" amusing; which makes it even more unlikely that they would be as easily fooled by the "weather balloon" reference as the professional skeptics who still regularly invent such preposterous explanations to dismiss UFO reports for the benefit and peace of mind of academic scientists, professionally addicted to elegant short-cuts.

^ ^ ^

As we summarize the information so far, we should recall that Sr. Anaya had come into town as part of the crew herding the sheep to the stockyards. He slept in the storage house and was awakened by a luminous object outside (what was it, by the way?) and by some "critters" in his room, who scurried away when he pulled out his rifle.

Jose recalls the sheepherder telling his father about "*three little ugly guys, about three feet tall, who started to put things into his mind.*"

Both military records and technical reports have long avoided the subject of the strange entities described by witnesses in connection with UFO encounters. When they mentioned them at all, it was in the context of unspecified "psychological effects," witness drunkenness or misidentifications. It wouldn't be until the early 1960s that write-ups of witness testimony would seriously address the paranormal issue, but then we always found them listed in connection with a psychiatric examination, or the suspicion of drug use. When psychic effects were involved, they were dismissed by professional psychologists as the simple impact of stress or as vague "dissociative" effects: mere short-cuts, unsupported by bedside interviews of patients or regular medical testing, a mockery of science.

Thus, to find these connections in the 1945 crash near Trinity is remarkable: Of course, there had been stories of Men from Mars and flying Aliens in comic books, in the "pulp" stories and in the early science-fiction of Edgar Rice Burroughs, but there was no mainstream literature at the time in the United States about UFOs, even vaguely suggesting psychic effects. Charles Fort did mention anomalous meteors, bolides and apparently conscious sky phenomena, but they fell short of the incident at San Antonio, where it's unlikely any of that literature was circulating.

Yet Reme Baca would testify: "The creatures were like us – children, not dangerous. I could see little creatures walking around, moving around, they either had grey-white coveralls, or they had overalls on and had a hood or maybe no hair, it was hard to distinguish. If they were wearing coveralls, they were skin-tight."

Reme had insisted: "They were like, staggering around, couldn't seem to hold their balance, all three of them. *They seemed to slide or glide from one end to the other* (…) there were sounds that continuously came from the object. High-pitched sounds, like the cry of a wounded jack-rabbit, sounds almost like a newborn baby crying, very moving."

∧ ∧ ∧

Chapter Eleven - Back in the Lab

Let's also keep in mind that in that 1945 San Antonio case, none of the people involved (the witnesses, their family or the military) used any of the terms or memes we commonly associate with the phenomenon today.

Even the term "flying saucer" wasn't used yet.

It would be another two years before Kenneth Arnold would launch the "flying saucer" literature, almost ten years before the Air force invented the term of "UFOs" and almost twenty years (1964) before Dr. J. Allen Hynek agreed to consider "occupant" cases within his casebook, at my insistent urging based on the French data and his direct experience with the Socorro landing.

In their effort to make the whole phenomenon seem rational, or simply comprehensible to their audience, many authors have glossed over the peculiarities, the contradictions in the testimony of witnesses who were simply trying to express their puzzlement or their terror when faced with entities that seemed able to place images in their minds or communicate transcendent concepts. Most of the civilian researchers actively involved in this research today still refuse to acknowledge the "psychic" angle, in their (misplaced) efforts to be taken seriously by the public, by Government and by incredulous physical scientists.

There is even an interesting dissonance between Reme and Jose on this point. Reme's testimony includes a mention of *"distinctly non-human* life forms which were alive and moving about upon our arrival minutes after the crash." Jose, however, insists on calling the creatures "men," as we see in his dialogues with Paola and with me, although in some of the transcripts he compares them to "fire ants" (Fig. 23). Yet Reme was quite precise. He said "They had big bulgy eyes...*you didn't even notice the nose because the eyes were so big*. They were thinly, you know, thinly means skinny, it was thinly built."

Somewhere else he tells Paola: "Everybody calls them Greys, I guess, but I haven't seen a grey, so I wouldn't know." His name for them is "Campamocha," Spanish for a Praying Mantis. He also simply calls them "strange-looking creatures," moving around inside.

Jose used another Spanish name for them: "Niño de la Tierra," an insect known as a "Jerusalem cricket" in English. He said they had four fingers.

"They looked under stress. They moved fast, as if they were able to *will* themselves from one position to another in an instant. They were

Trinity: The Best-Kept Secret

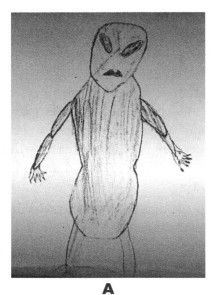

A

A: UFO occupant described by teenage students to Dr. John Mack in Zimbabwe, 1994

B

B: "Jerusalem cricket," approximation to what Reme and Jose saw. (Note: The insect, also called Niño de la Tierra, is shown here without two of its "arms" for comparison.)

C

C: "Praying Mantis," "Campamocha" (Reme)

D

D: "Fire ants, standing up" (Jose)

Fig. 23: Appearance and terminology in frequent, reliable accounts of "entities."

shadowy and expressionless, but definitely living things. I wanted no part of whoever, whatever was inside." (**12**, p.38)

Yet on the way back, riding his horse and helplessly crying, Reme couldn't push away the experience, and he too had some visions he recalled 65 years later:

"*I saw these pictures in my head...*" he told Paola in 2010. (**12**, p.113), "but I didn't know what the heck they were. I can remember what they are, *I got pictures*, but I didn't know what they meant by then, and I still don't know."

Dr. Jeff Kripal of Rice University has pointed out that the Alien in modern abductee accounts is often described as insectoid:

> "These haunting figures are frequently compared with an immense praying mantis or human-sized insect. They communicate with humans telepathically. The immense almond-shaped eyes are compared with those of bugs...The modern abduction visions thus replicate in precise and eerie detail, the thought of Frederic Myers: the same perfect insectoid form (Myers' imaginal stage), the same telepathic powers (Myers coined the term)... How do we explain such precise resonances between this Victorian psychical researcher and our contemporary abduction literature?" (**30**).

Professor Kripal's observation is very apt in this case. It applies well to the children's impressions in 1945 San Antonio, except for the fact that—to the difference with abductees whose testimony we regularly hear on television today--they were fully conscious at the time, had never been subjected to hypnosis, were not victims of an abduction real or imagined, and cannot be suspected of being influenced by medieval notions about Magonia or by esoteric Victorian literature. And they experienced their ordeal many years before the topic of abductions, or the very expression "flying saucer," was ever taken into account by researchers.

∧ ∧ ∧

CHAPTER TWELVE

A TRINITY OF SECRETS

Everything, in this story, appears to be going in threes. It makes it easy to remember details.

Three atomic bombs were exploded in the summer of 1945, putting an end to the Second World War.
- One at Trinity
- One at Hiroshima
- One at Nagasaki

There were three live *Campamochas* aboard the crashed craft recovered by the Army, although they may or may not have captured them, or even seen them, since their reports are vague.

Separately from those, there were three "short ugly guys" who "*started to put things into the mind*" of the sheepherder, who wasn't aware of any of the ongoing events.

Also, at least three metallic artifacts were recovered: The bracket itself, the piece of foil that was used to fix a windmill, and an I-beam that was disposed of as scrap metal. The only one for which unusual properties were described was the foil, because it was supposed to regain its original shape when folded, like some of the material that reportedly came from the Roswell crash site.

Unfortunately, there is no way to test that claim. The only object we have been able to test was the bracket, and it turned out to be a common piece of human technology, as far as we know so far...

In fact, the only strange thing about the bracket is why so many human and non-human creatures seem to be eager to recover it – and hide it!

^ ^ ^

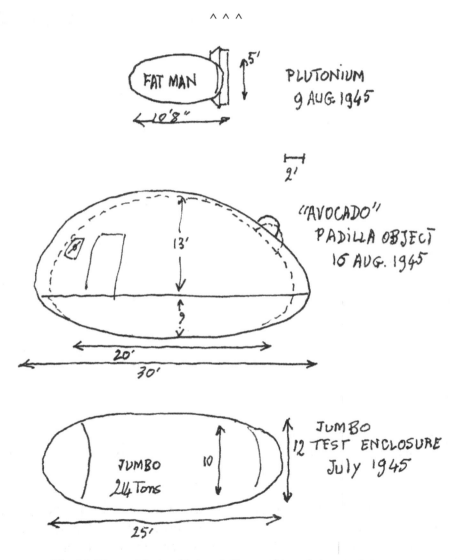

Fig. 24. Three objects of interest: Comparison of sizes and shapes

Chapter Twelve - A Trinity of Secrets

When you look at the case in that light, as the result of serious, converging investigation by multiple teams, a lot of facile, convenient hypotheses or explanations evaporate.

First, we have to ask a question that seems preposterous, or at least bizarre: What if the object was *designed to crash*? What if the lives of the creatures aboard were expendable?

Absurd? Perhaps. But look at what the awesome burden of responsibility the brightest minds in the United States had just taken, in the name of science and in the pursuit of war: Many thousands of civilians dead in Hiroshima within seconds of the blast; as many in Nagasaki; long-term illnesses for many of the survivors, and the horrible prospect for large-scale deployment of the new weapons in future wars. The Soviet Union would soon have its own bomb, followed by the British, the French, the Chinese, Israel and even some countries that are permanently on the brink of conflict, such as India and Pakistan. It seemed that the appetite for what would eventually be called "weapons of mass destruction" was irresistible even when it became generally accepted that the victors would not survive much longer than the vanquished.

Nor was it possible to claim ignorance of far-away impacts on poorly-documented populations: The effects of the bomb could be readily seen at home, right in New Mexico. In the name of secrecy, military authorities had not wanted to relocate or protect the local population, but they knew roughly what the likely fallout from the "Trinity Test" would be: the responsibility was dreadful and the price in human misery was enormous.

The American Heritage Dictionary defines a euphemism as "*the use of a mild, delicate or indirect word or expression in place of a plainer and more accurate one, which, by reason of its meaning or its associations or suggestions, might be offensive, unpleasant or embarrassing.*"

Calling what happened at Ground Zero a "test" constitutes a nice scientific euphemism, because the power of the bomb was reliably estimated at 19 kilotons, which places it between Hiroshima (12 to 18 kilotons) and Nagasaki (21 kilotons).

The Nagasaki device was the same type of bomb as the Gadget, but it was detonated at 1890 feet to avoid long-term radioactive contamination of the area, as opposed to the explosion at White Sands, which took place at ground level, with the worst possibly effect on the environment. Yet historians, even today, blindly copying the Pentagon verbiage, continue to daintily refer to the Ground Zero explosion as a "test."

Remigio Baca recalls a neighbor named Cecelia talking about her farm animals struck with an unknown disease: "There's a kind of mucus coming out of their beaks; some of the cows lost their hair." Another woman, Ana Lee, said her sisters developed cysts, and she had three miscarriages. *"We didn't find out the radiation dangers until after the war; they should have notified us about the risks."* (**12** p.15)

Reme Baca's book provides a sobering, overall review of the impact of the Trinity event, quoting government documents that leave no doubt about the level of plutonium contamination and the size of the affected area in New Mexico. A 1983 field study noted there were large areas near Ground Zero where vegetation was not growing, even after 38 years. That was one of the things that struck us on our long drive to Stallion Gate: Where were the trees, the bushes, the abundant cactus that enlivened the old descriptions from pre-1945 days at the Padilla ranch? All we saw were thousands of acres of low-growth weeds and short bushes along the road to Ground Zero. The zone of the crash site itself remains quite bare today.

To be fair, soon after the review of the fallout, Army documents recommended that future tests be held at least 150 miles from populated areas, which led to the permanent relocation of the test site to Nevada. But that decision came too late for the Padilla home, which was less than 20 miles northwest of Trinity. For the first half-hour the fallout from the bomb had drifted to the northeast, but the winds then shifted to the northwest, the area of San Antonito, depositing fully one-quarter of Trinity's radioactive debris on a region extending to the Chupadera Mesa.

Radioactive debris rained over the area for days.

If visitors from elsewhere had wanted to sample the effects of the first atomic explosion one month after the Trinity event, they certainly could not have found a more appropriate location anywhere on the Earth.

To the credit of the US military, the terrible impact of the "test" at Ground Zero was noted, investigated and reported. (Note)

Studying the documents from the time, I learned a stunning "detail" that had completely escaped me: the purpose of the bomb was never to spread radioactivity or its fallout. Instead, it was massive, instantaneous

---

Note: The destructive power of atomic bombs used at Trinity and in Japan was measured in *kilotons*. A few years later, with the development of Hydrogen bombs, the explosions were one thousand times more powerful and became measured in *megatons* of equivalent TNT.

Chapter Twelve - A Trinity of Secrets

destruction by enormous air displacement and extreme temperature: essentially, a giant, deadly firestorm. So, based on the rushed measurements in the area of San Antonio where much of the radioactive debris had fallen, the scientists recalibrated the Hiroshima and the Nagasaki devices so they would explode at that higher altitude, about 2,000 feet, sparing the Japanese people even greater misery and making it possible, in time, to rebuild the cities that the war had pulverized.

As I recalled my own trips to Japan, and the business meetings I'd attended with financial executives in areas that had been targets during the war, I couldn't help thinking that whoever made that decision at Ground Zero deserved some sort of discreet medal. But I don't expect Tokyo will send him a branch of flowering cherry tree any time soon.

∧ ∧ ∧

Let's go back to our secluded laboratory with all the charts, maps and calculations pinned to the walls. One of the things we still need to do, with no preconceived notion about the events, is to reconstruct a model of the object from the various testimonies of the main witnesses. Since we have no access to the records of the military task force, we can only use Reme and Jose as our sources, but that happens to be very good because they were on-site for a full seven days (including the day of the crash itself) and they could see the object from different angles during the recovery process.

As we have seen, they were able to approach it and touch it, inside and out.

What they describe (contrary to the rumors that would start later, and to various overly-imaginative drawings in UFO magazines) it was not a disk or a "flying saucer" at all. It was, as Jose puts it, something resembling the shape of a flattened avocado, 25 feet long (plus or minus five feet) and not extremely heavy.

In our follow-up to the first series of interviews, Paola and I probed Mr. Padilla's recollection of the loading of the object onto the low-boy trailer. Specifically, how many axles, how many wheels on that trailer? He clearly recalled a military 18-wheeler with four axles and a tractor in front, and a 35-foot platform that could be lowered to the ground to pick up heavy loads. His rough guess at the weight of the object itself was less than 10,000 pounds, the low end of the range for such a truck. Note that

"Little Boy" weighed 9,600 pounds and Fat Man 10,300 pounds, in the same range.

This gives us an idea of the values for the physical parameters we need: Essentially, this was a heavy object the size of a big van and about the weight of two Ford F-150s. No wonder nine-year-old Jose Padilla couldn't shake it when he tried!

∧ ∧ ∧

When we assemble all the bits of testimony, we end up with an object with the shape shown on figure 24: A bulbous, roundish vehicle, grey or dark in color, built out of an unknown metal with no visible rivets or screws joining the various panels together, a flat floor and very high ceiling, about 13 to 15 feet. And no visible means of propulsion. Presumably there is some space under the floor but we don't know if it holds anything of interest, like what Government physicists would think of as an engine.

The only other item in the shape is the very small transparent dome, off-center.

It is also extremely strong, having sustained a crash and catastrophic slide over rough terrain the length of a football field without breaking up. One panel has cracked open but when Jose Padilla examines the underside on the last day (the object is laid down sideways on the trailer, because otherwise the load would not have been able to drive under the highway underpass on the way to Stallion Gate) he only sees scratches and deep abrasions on the surface but no break. There are no openings, and no appendages.

Let's note in passing that any aircraft of the time would have broken up into many pieces if subjected to the same catastrophic crash at that location. The fact that the "avocado" was able to keep its integrity (without tumbling) under the circumstances, even after destroying a radio tower on the way down, tells us something about its remarkable construction. That, itself, would have warranted the attention of the classified engineering contractors who were later asked to analyze the crash.

In 1945, the only way to assemble large metal plates involved visible rivets.

Next, we have to ask if the object, as a physical device but also as a symbolic "signal," resembles anything in its environment at the time, and two connections come to mind immediately: The atom bomb itself was shaped like an avocado, and our old friend "Jumbo" is an elongated cylinder with rounded ends. Having examined both of these, on-site at Ground Zero, Paola and I have been struck by the similarities.

Shown together on figure 24, the three shapes invite serious reflection.

Jumbo, in particular, has several characteristics in common with the crashed object. It has very strong integrity and similar dimensions: 25 feet long and 12 feet in external diameter. Like the crashed device, it is empty. One major difference is the weight: Jumbo weighs 214 tons, versus our rough estimate of about five to ten tons for the Padilla object.

On figure 15, the picture I took of the packaging of "Fat Man," a comparable avocado shape is again evident, although of course the size is smaller but the weight is in a similar range: It had a mass of 4.7 tons, which puts it close to "our" avocado.

Was an external agent trying to tell us something? What concerns did it raise in the minds of the atomic scientists? Were they even briefed about the intrusion? **(31)**

...and the most important question:

Has the conversation continued?

∧ ∧ ∧

As we proceeded down this road, and sought advice from scientists and historians of that period, we naturally were greeted by skeptical views, *which we welcomed.*

Much of skepticism is short-sighted, born of prejudices and hasty judgment, antagonistic to the point of dishonesty and generally useless, but there is also a positive meaning to the word, directly derived from its Greek roots in philosophy. In that vein, the school of skepticism teaches that one should consider conventional phenomena, and exhaust the research into subtle causes derived from established knowledge, before rushing into new theories and fantastic claims. We were eager to use that "good" skepticism to clarify the landscape before us.

One persistent argument against the "Alien" nature of crashes, and especially the post-war UFO crashes in the United States, has to do with the possibility of Soviet interference and even, in an extreme hypothesis, of surviving Nazi underground projects involving sabotage, confusion or intimidation on American soil. Throughout the Manhattan project, naturally, the US Army was on high alert, and it was known that both the

Russians and the Germans suspected radical new developments at White Sands and Los Alamos, and tried hard to interfere and infiltrate.

This sent us back to the state of international knowledge about secret weapons, real or imagined, and their development up to 1945: Could the craft seen by Reme and Jose be a device invented by an enemy, deployed either as an intelligence-gathering tool or simply as an intimidation tool or as a warning?

We have to admit that the date, two days after the capitulation of Japan, assured that whoever crashed that craft on the Padilla land would get full attention from the US Army.

What is the logical argument against the event being organized and managed by human forces? There is a vast literature about German secret weapons, Soviet Intelligence and the state of the art in aerospace research, yet it provides nothing resembling what our four primary witnesses (Jose, Reme, Faustino and Apodaca) had seen during the crash or within hours of it. Setting aside the fact that it would be foolish to expose an experimental prototype and give it away to the US if you were an outright enemy or a competing nation, we will see in the following chapters that the technology involved was even more extraordinary than the kids could imagine. In fact, we have nothing of the kind even today.

A declassified American Intelligence document entitled "Evaluation of German Capabilities" reviewed "the actual or potential weapons which the Germans may use against USSTAF Operations in 1945," setting aside "those for which there is lacking evidence of possible use for time to come" and concentrating on rocketry, from the V-1, the V-2 but also including "Phoo bombs," magnetic waves, "Gases applicable to aircraft" and even the German Atomic Bomb, considered unlikely to be operational in the time frame under consideration.

An additional report of interest at the time concentrated on "Engine interference by electro-magnetic disturbances," which naturally has relevance in many modern UFO reports by both civilians and the military. But again, none of that relates to the extraordinary vehicle seen at San Antonito. We had to leave that search aside and look at other sources of confusion or error.

In a very comprehensive book entitled *Man-made UFOs, 1944-1994*, Renato Vesco and David Hatcher Childress also review the full range of secret vehicles conceived, built and tested by Nazi Germany towards the end of World War Two. Some of them were remarkable in the way they anticipated modern weapon platforms, but here again, one should not confuse an interesting experiment with a fully-developed flying craft. The designs that reached some level of test in factories and airfields from Spandau to Prague in the 1943-44 time frame, apart from the Feuerball and Kugelblitz mentioned above, belonged in two families of craft: circular saucers equipped with turbojets or gas turbines, and advanced rockets or rocket planes.

In his introduction to this very well-documented book (at the price of some confusion) W.A. Harbinson, author of the novel *Genesis*, remarks, citing Vesco: "the *Feuerball* was a flat, circular flying machine, powered by a special turbojet engine, which was used by the Germans during the closing stages of the war both as an anti-radar device and as a "psychological" weapon against human pilots."

While that reference is intriguing at the dawn of the modern flying saucer saga, let's remember that all these devices had conventional propulsion by modern standards and did not resemble what Reme Baca and Jose Padilla observed. No matter how far I tried to follow the authors' complex research, I have not found anything resembling the craft that crashed in San Antonito, which certainly didn't have conventional engines and was neither a disk nor a rocket.

Pursuing our investigation, I wondered about the beings aboard the craft described by the two children. I tracked down an interesting old military report that mapped "Anthropomorphic Dummy Launch and Landing Locations" in the White Sands area. (Note)

Here was another series of very real projects that had to do with experiments on future devices to be used by humans in the high atmosphere or in space, including special flight suits and parachutes. The test articles were humanoid dummies approximating real bodies in size and weight. They were primarily launched, according to one map, from the San Andres Mountains directly West of White Sands, and were recovered after short balloon or rocket flights all the way from the Jordana Test Range to the Sacramento Mountains and even Roswell.

Although the implications are intriguing, since some of these devices and their "pilots" would fall in the area of San Antonio, they are useless for our purpose because the period is all wrong. Let's remember that the Air Force didn't even exist at the time of the Padilla crash and that the US

Note: *Air Force Aeromedical project no.7218*

Army did not have sophisticated manned balloons or rockets until later, when large amounts of German equipment were moved to the test sites in the Southwest. Furthermore, the kids had seen creatures that were alive and could not be confused with the humanoid dummies.

Although the parallels are interesting historically, they bear no explanatory value in the case of Jose Padilla: the anthropomorphic dummies were simply hanging from parachutes dropped from balloons and no sophisticated craft was involved in the launch or the recovery.

From a practical point of view, let's also remember that those projects were not begun until *Project Paperclip* had resettled German engineers and their rockets to bases in the Southwest: Wernher von Braun launched the first V-2 from White Sands in the spring of 1946. Recall also that space research was long a topic of ridicule in the United States. A remarkable book called *The Pre-Astronauts*, by Craig Ryan (Naval Institute Press, 1995) notes: "in the late forties through the mid-fifties, "space" remained a dirty word in the Pentagon and in the halls of Congress. To mention the subject was to invite jeers."

Sort of like the word "UFO" today...

∧ ∧ ∧

Satisfied that there was no evidence for any human technology of the time approximating what Jose Padilla had seen, we went back to the details of the case. Once we had clarified the circumstances of the event, with a tentative model of the object that resolved the various interpretations and obvious errors in some earlier, fanciful descriptions, Paola and I found ourselves before another, more subtle series of open questions. Having reconstructed the sequence of events, both from the Army's standpoint and from the local reality of the site, a number of new interrogations emerged.

The first question I had was about the matter of the Marconi tower, since we had established that the craft had hit it on the way to the crash site.

"The map is not the territory," as Alfred Korzybski famously said, and you don't need General Semantics in the middle of a desert where weather and terrain conditions can change radically around you in less than half a day. Something as simple as reconstructing the position of the northern radio tower for the White Sands range had become an interesting puzzle, but of course it was critical to the trajectory of our "avocado."

Chapter Twelve - A Trinity of Secrets

The conversation began with Paola saying: "You gave me a 1982 map, which is beautiful."

Jose: Ah, the map... I put everything there.
Paola: All right, you put the Padilla house, you put the well, you put a lot of things on there.
    And the tower! The tower is interesting where it is. Is that the Marconi Tower?
Jose: The tower was on a BLM map. The one who made the map, at BLM, is the one that gave me that letter for a permit to go in there.
Paola: But the tower isn't there anymore, so how did she put that in there?
Jose: It was there. She put everything that... I can't remember her name, she's retired already. She's the one that wrote that, all the map, and then I added all the little things in there, like Diamond-A Ranch and the windmill, and Walnut Grove.
Paola: Yeah I understand that, but the tower... If the map was in 1982, the tower was destroyed by then. Why is the tower on the map?
Jose: The tower wasn't there no longer, it was destroyed. The tower was taken out of there, way after, let's see, 19, uh, 45, 50s -- 56.
Paola: The tower was taken out of there in 1956, but the map is from 1982. It's like 25 years later. Why is the tower still on the map? Did you put the tower there?
Jose: No, no, the girl from the BLM put the tower there because she knew that there was a tower there, she put it down.
Paola: So what we have is a problem. We have to look at the trajectory of how that thing came in. It took out the tower and landed at the spot where you have the X. Where did it come in from? Over your head? Did it come in behind you? And then went and landed there? I mean, where, what direction did that craft come from?
Jose: The direction the craft came from, was from San Antonio, like, coming from that Trinity Site, and then it made a turn and when it hit the tower, that turned it backwards and it landed right where it crashed.
Paola: OK, so you think the craft came from the Trinity Site?
Jose: It was my feeling, that it came from Trinity.

^ ^ ^

Trinity: The Best-Kept Secret

The next remaining puzzle was equally important. It had to do with the circumstances under which the tower was taken down, as well as the gate the Army had installed back in 1945 to drive in their big truck. It also had to do with one of the metal samples recovered at the site.

So Paola inquired about the time when the Army came back.

Jose: When was it? Was it 1956 or '57 – that's when my cousin was going out there and American Police stopped him at gunpoint. He couldn't go in there because it was Federal Property and they wouldn't let him go in there. At that time it was, uh, '56 or '57 when they tore that whole thing down. The gate, the tower, and they demolished it.

Paola: They tore the gate down, too?

Jose: Yeah, they took the gate, they took that tower down. That is when my cousin was strolling on his horse and he found a piece of aluminum, out there in the middle of the field. About four, four miles uh, southwest, of where that crash was.

Paola: You mean the I-beam, the I-beam piece, yes?

Jose: Yes, the I-beam.

Paola: Yeah, ok, so in that year, you said 1956, the Federal government came and took away every evidence that they had done anything?

Jose: Yes.

Paola: They took away the gate that they put there, and they took away the tower that was destroyed?

Jose: Well, the people took the gate down and took it and sold it as scrap metal.

Paola: So what were the soldiers doing there?

Jose: *They were looking for something.* My cousin and a friend of mine were going up there to check and the military police pulled them over before, before that cattle guard, and told them, you know, they couldn't go in there. That's when they tore down that gate, the pipe gate. They tore the tower and uh- they demolished it, and that was it.

∧ ∧ ∧

Little by little, the drifting pieces of history were floating back in the current of time, through that remarkable memory of Mr. Padilla, and the collective memory of that little isolated part of New Mexico.

Chapter Twelve - A Trinity of Secrets

The next, incomplete part of the story had to do with the integrity of the craft itself, when it came to rest against the dam in the arroyo. Paola relayed my question about it.

Paola: Jacques is wondering if that panel that blew out, and was damaged, was it a door? If it was not a door, if it was just a panel, then how would those beings get inside?
Jose: Well, that piece was never found; that part that popped out of, out of that crash. When it hit that angle of that tower, that piece popped out. I don't know whether that was a door, an emergency door, or what. But that was never found.
Paola: OK, so if that was a craft that had no door, it was just a panel. Do you think it was a panel or do you think that was a door?
Jose: To me, it seemed like it was a door.
Paola: It looked like a door because it was clean? Was it a nice clean panel, or was it all jagged?
Jose: It was a panel that didn't have no hinges on. Or possibly it had some type of adaptors in that, a pop out thing, like an emergency outlet.
Paola: So it was not like a piece that was broken off. You didn't see pieces all over the place, right? I mean, was it broken or did it just pop out? How did, I mean... 'cause if that was part of the craft, then there is no door, and *how did the beings get in there?*
Jose: To me it, uh, it was like a door that just popped out.
Paola: Because... why? Was it clean? Was it nice and clean? I mean, was the panel on the ground?
Jose: There were some... some dents where it hit the corners of that tower.
Paola: There were dents on it?
Jose: Yes, where it hit, uh, the corners of the metal of the tower. And I don't know whether there was suction, or some kind of a thing, that made that thing pop out.
Paola: Because people are going to ask, how did the beings get in there? Did they just go through the wall? How did they get in there – you think it was a door.
Jose: Well, there was a way that, if it belonged to them, you know, there is a way that they can get in there, if they wanted to get in there or out.
Paola: Yeah, if they could go through the wall, the way that they did in the sheepherder's shack.

Jose: That's what I was going to tell you. That is they, they get into a wall out there, I mean they can go and get in, in anywhere. Now, in a way, you know, I wish I could've gone there and touched them to see what they were made out of.

Paola: You think that panel was a door, then. You think it could have been a door, or it was just another panel, the way the panels were inside, right?

Jose: Inside in there, it seemed like there were panels, but only that section, right there, is the one that popped out, and it was never found. The only thing that was found is that piece of metal that my cousin found on the field when he was going home on the horse. But he never found anything (else), up to now, and *he still doesn't believe in any UFOs*. So he didn't care. He thought it was a piece of metal. He threw it out on that trash pile, to sell as scrap metal.

∧ ∧ ∧

That left the question of what the panel actually looked like, and what had happened to it. The best hypothesis, as we saw before, was that a small Army team, alerted by the pilot's observations of the smoke and the crash, came in on the second day (when the kids couldn't come to watch) and took away some of the material.

Paola asked: How big do you think the panel was?
Jose: Uh, the panel was about, uh, 4 feet by almost, almost 9 feet.
Paola: So you think somebody went out there and picked up the panel?
Jose: Someone picked up the panel and took it.
Paola: OK, and when you and Reme went down there, did you see the tower destroyed already?
Jose: One corner was. Up above, about three quarters of the way up. It was a quarter of it, it was all mangled up.
Paola: So you knew this was a real thing. 'Cause you heard the boom, and then you saw the mangled up part in the tower and then you looked down and saw what did it. Is that right? You guys heard the boom, then you saw the mangled part of the tower and then you saw what did the damage, you saw it.
Jose: The mangled part of the tower was northwest.
Paola: OK, but you know that the craft did it? You know that the UFO did it?

Chapter Twelve - A Trinity of Secrets

Jose: That's the only thing that I could figure out that it was, you know, because the parts that were all scattered were part of that. I will call it angel's hair, whatever it was, that was on the ground there. Because it was a pretty big space where that thing came, from where it hit, all the way to down to where it landed. That is a long ways. And there was a lot of things scattered all over the place.

Paola: So the angel hair stuff was scattered all over the place?

Jose: It was scattered like, when something blows up, it blows for a long ways, you know. That is why that piece of material that my cousin found was four miles away south from there. To me, it seemed like it was some kind of insulation, between the panels.

Paola: OK, but was that near the tower? Or was that on the crash site?

Jose: Some of them were by the tower there, where it crashed, and uh, some of the stuff, most of it, was on the ground where it landed to rest.

Paola: That's so important Jose, that detail, I can see that.

Now we could reconstruct a better model of the collision with the tower, and that was important to understand the actual damage to the craft, which seemed limited to that single panel. But we still couldn't account for the fact that the "door" itself was missing, so the dialogue continued:

Paola: When you were looking through the binoculars, did you see that door panel, or the piece on the ground – did you see it, or did you...

Jose: No, no, that was blown out. Well, you know, how far it was from that tower to where it landed? That was blown out when, when it hit the tower and um, we don't know where it blew out to.

Paola: So you really never saw it?

Jose: No, no, we never saw that. The only piece, you know that my cousin found was possibly a piece from that. But that was four miles away from there.

Paola: Ok, so, in other words you never saw the panel, the door, we're calling it the door, but it's not necessarily a door, it was just a panel. But you never saw it, right?

Jose: No, I didn't. It was just, uh, popped out, I guess, from that impact...

Paola: Then you said that you think the panel may have been buried under something because nobody found it. Maybe it's buried under dirt.

Jose: Yeah, it could be, you know. The ground grows. It grows up fast within years and uh, my cousin was lucky that he found that piece of metal that was lying up. He was going through there, uh, the winds will have blown a lot of things away and blowed dust, and then he discovered that, on top of it.

∧ ∧ ∧

It may seem that we were drilling endlessly with such questions, but it's not every day that you have an actual witness of a UFO crash, who is so skilled at recalling the details of the disaster. So we centered our questions on the other material retrieved from the site, namely the "angel hair," and the exact trajectory of the craft before it hit the ground. In the following weeks and months, we would have lots of questions about that "angel hair."

Paola, in daily contact with Mr. Padilla, first asked about the material he'd picked up:

Paola: About the "angel hair," do you figure that it came from that panel, and it was at the tower?
Jose: I have a feeling that that "angel hair" came out of that (panel), like insulation.
Paola: You said it was white, like a web? Like strings – was it strings?
Jose: It was white, uh, clear white, like a web, but, in a lot of little pieces, you know, that you could blow, like insulation.
Paola: But was it like strings? Was it strings?
Jose: It was strange. It's the first time I ever seen it.
Paola: But was it strings... Or was it more like a web?
Jose: A web. Like a web, um, you can blow it and you'll blow like a flower, when you blow on a flower and the leaves fall off.
Paola: So you and Reme actually picked up that stuff?
Jose: Yes we did. One thing that we did try to do is burn it.
Paola: But you touched it – was it light?
Jose: We touched it and it was fine, like a web, you know, it was a fine material. That's why I have a feeling that, all that, angel hair or whatever, or insulation, that when it hit so hard, it just exploded and disintegrated and just blew away, you know.

Naturally, it occurred to Paola that we might be able to recover some of that material: It would be interesting to do a thorough analysis of its components.

Chapter Twelve - A Trinity of Secrets

Paola: How much was there? Like, a lot?

Jose: We accumulated quite a bit of it. We put it in a sack.

Paola: You put it in a sack?

Jose: Yeah, and we used to have decorations for Christmas and make it out of that...

Paola (*laughing*): You're kidding me! You used that angel hair in the sack for Christmas decorations, and maybe that came from another planet?

Jose: I guess, you know, we, we done it! (*chuckles*)

Paola: My God, are you kidding me, Jose? (*laughing*) You know what you are saying? You're saying that you took all that insulation from the inside of a space ship and made Christmas decorations, like, you really did use that stuff?

Jose: A lot of that, you know, that we found on the ground, we'd put it in a sack, and we took it home and we gave some of them away. And the people around the community there, they used it on the windows of the homes. They'd look like a spider web....

Paola: How many bags did you collect of the angel hair, do you think?

Jose: We had the hundred pound bean bag, where we used to put beans. Well, it was that size of a bag. It's actually like the size of a 55 gallon drum.

Paola: But what did you guys do? I mean, were you walking around just picking it up, putting it in there?

Jose: We just picked it up and took it up there, and we decided later it would be a good... We didn't have enough cotton to make angel hair to put on the windows and we put that on, and it looked good.

Paola: (*laughs*) Jose, so you never saw that stuff, except when you picked it up around the week of August 16th, right? You never saw that anywhere?

Jose: Ah, no. You know, a lot of the people that used to see it, used to ask us where we got it. We told them that we just found it. And that was it. You know, it was a secret.

Paola: How many people did you give it to?

Jose: Some of them are our neighbors, and some ranchers.

Paola: OK, and then....

Jose: One thing that, um, that *we tried, was to burn it and it wouldn't burn.*

Paola: What made you think of burning it?

Jose: Well, to try; it wouldn't.

Paola: You just were doing, you wanted to see if it burned, and it wouldn't set afire.

Jose: Like, cotton, you know, you can, uh, uh light a match to it, and it'll burn. And that stuff didn't burn at all. (Note)

Paola: It didn't burn; what color was it again? You said a clear white?

Jose: It was a clear white, like a spider web. You know when a spider web gets dusty? That's the way the color was. During the Christmas season they'd seen that we'd decorated our homes with that on the windows, and the people that were going by, you know, they'd seen that, and they stop and ask, what was it? You know... it's a weird thing. It looks like a spider web. At that time, we used to use cotton. And they noticed that it wasn't cotton. They asked where we'd picked it up. We didn't say where we picked it up, you know, but we gave some to people, and, they decorated their windows with it.

Paola: But Jose, the crash was in August... did you put it in the garage until December?

Jose: No, we had sacks that we put it in, and we just, uh, put it on storage in the barracks.

Paola: What do you mean storage in the barracks?

Jose: At the Ranch we had cabins, where the ranchers used to sleep and then get up next morning, you know, and we called'm barracks. And we put them out on bags in there, and just set them right in there.

Paola: In other words you didn't use it right away; you waited at least six months.

Jose: Just for the Christmas Season, we used that and it was weird, like, cotton that we used before, we could burn it. This stuff here, we would light a match to it and it won't burn.

Paola: It wouldn't burn at all, so you light a match to it and ...What made you think of lighting a match to it?

Jose: To see, see if it, would blow up or whatever, you know.

Paola: Because you're little kids, and you want to see if this stuff is gonna blow up!

Jose: Uh, huh, I just wanted to, I was curious, what that stuff was made out of, you know? If it would burn or not...

Note: This is important, because a very early prototype of optical fiber would have been destroyed by the heat of a torch. So the material was not fiberoptics as we know it.

Chapter Twelve - A Trinity of Secrets

Paola: But if that craft hit the tower, wouldn't most of the angel hair be around the tower?

Jose: No, most of it blew away. With the impact, that panel must have disintegrated, that angel hair blew all over the place, and it used to hang on into the bushes and on the ground.

∧ ∧ ∧

When Alfred Korzybski stated (in his *Second Principle of General Semantics*) that "The map is not the territory," he must have been thinking of New Mexico, I thought as I drove up the dirt road one more time. As another piece of the puzzle, we needed to pin down more precisely the trajectory of the craft before it crashed, which meant reconstructing how it collided with the Marconi tower, but we didn't know where the tower was, and the maps were of no help.

In the course of the inquiry it had become clear to us, not only that "the map is not the territory," but that some of the things marked on the map were only the vague souvenir or departed spirit of structures that had existed once, but were no longer in existence on the land itself after 1980.

In order to wrap up that phase of our investigation, we needed to locate the tower with greater precision. If the local maps suffered from a bit of fevered imagination on the part of the cartographer, then only someone who had grown up on the property could tell us what had actually been there.

Paola asked, "We have the map you gave us, and it shows a tower. It's really not near the gate where you come in. So, were there two towers?"

Jose: Um, no, no, no.

Paola: Well, where (the cartographer) put the tower on the map that we have, is not near the gate or the cattle guard. She put it somewhere else, up north.

Jose: Well, no, when she drew that map to give it to us in 1980-something, 1983– or whatever, she just put the tower there to make sure that there was a tower, you know.

Paola: So it's not necessarily accurate, is it? It's just a placeholder?

Jose: No. That tower was... remember when we found a little piece of wire?

Paola: Yeah, it's near the cattle guard. And so what Jacques was wondering is, when the craft hit the tower, did it fly downwards, or did it go back up again, 'cause he's looking at the elevation. It looks like it went back up again (*if we're talking about the tower up north*-JV).

Jose: Uh, no, not, quite, because that is downhill. From that tower it's all going south and everything back there is almost downhill, down to where the crash was.

Paola: OK, because he's looking at topological maps and wondering. I said to him, "No, it skidded, you know, it skidded a long way." And Jacques said, "So, when it skidded, did it make a left-hand turn?"

Jose: Uh, when it skidded, let's see: It hit, going west, when it hit it went south....it did turn a left hand.

Paola: And that's when it stopped? Do you think it bumped into something, or it just stopped?

Jose: Um, it stopped, that was the end of it, because there was nowhere else so it could skid.

Paola: Oh, 'cause if it kept skidding, where would it skid over?...over the arroyo?

Jose: Um, no, see where that road – it was built? That's where it came down. *Like if they could have guided there.*

Paola: OK, so what you're saying is, the new road is almost the trajectory of the craft?

Jose: Right.

Paola: The tower was built on a hill. It's near the gate. And it was built as a signal so the planes wouldn't slam into, into what – the tower? or the hill?

Jose: Uh that, Remember the cliffs? They were on the opposite side? Where we took shelter? The tower was built there, so they could signal the airplanes 'cause they went through there, so they wouldn't hit that hill because that was the highest point there.

Paola: Yeah, OK, well, I got it. Maybe it wasn't a radio tower. Was it still a radio tower?

Jose: My Dad named the place "Pico de Italiano." That means *Italian Peak*, which is on the map there. And I don't know whether the military or whatever built that tower; the purpose of that, according to him, is that the aircraft that used to fly from the south, from El Paso, flying north, wouldn't hit that hill where the tower was mounted on.

Chapter Twelve - A Trinity of Secrets

Paola: Ok, well, Marconi built radio towers, so that's why they called it. Maybe it's a Marconi tower, which is Italian. So, you know then, when your Dad and the other guy went there, they certainly didn't find the door (*panel*). Nobody saw the door ever, right?

Jose: Uh, that door never did appear.

∧ ∧ ∧

Paola: So, it was on the north side of Walnut Creek. It's really hard to see the line of the creek there. But on Jacques' map it shows the crash site being a little bit farther away than the three miles, but I remember it was roughly three miles to get down there. So, when you drew the picture of the damage, how did you actually see the damage and how long was that tower there? The radio tower that the craft took out.

Jose: The one where the crash was?

Paola: Yeah, did that fall? Or did you just see the damage, and it was around for a while?

Jose: No, no! *I climbed up there.* I climbed up there, and I said that's where I put down, on the diagrams that I wrote there, that uh, the damage was made from the south-east. That's where it hit.

Paola (*astonished*): So, you actually, at nine years old...climbed that tower?

Jose: Yes I did.

Paola: What made you think it came from Trinity?

Jose: The way that I seen that, uh, the, the ricochet of where, how it hit. And it bounced south-west from there.

Paola: The ricochet. Interesting! Could you see the crash site when you climbed up there?

Jose: Uh, no. It was too far. Well, I didn't climb up all the way up there. But I climbed up... at seventy five feet high... I climbed about three quarters of it. That's where I checked where the damage was where it was coming, it was coming from the East, I mean what do you call it? *Trinity*. It was coming from there. When it hit, it hit the corner of the leg and it ricocheted away, the way the scratches were – it ricocheted and headed south-west.

Paola: Wow, that's amazing because that's a physical traces situation and you climbed it and saw it: saw the damage, so something had to damage that. How soon did you climb it after the crash?

Jose: Um, right after that, uh, that *apparatus* was moved from there.
Paola: In other words, after they took it away you climbed it, like what, a week or two, or?
Jose: Well, yeah. That's when we started finding this angel's hair that I was telling you about.
Paola: And did that picture that I sent you –did it look like that?
Jose: Uh, no. No that's completely, completely a different thing. I checked that and it wasn't like that. This was like a, a spider web, um, it's hard to explain. But it was like a spider web, that would usually gather anywhere, on the bushes or anything like that.

^ ^ ^

That wasn't the end of the conversation, because there was a military presence in that area for many years, and Paola was intrigued by some of the incidents that suggested ongoing search for traces of the incident. She asked: "You were saying when they pulled your cousin Ed over at gunpoint, what year was that?"

Jose: I believe that was 19 uh, 50....1956. Right there before going up the hill, before you hit down to the level where the tower was, before the cattle guard.
Paola: And what happened?
Jose: They turned back. They turned back and went home, and then they got on the horses and came out the back way; the way I go into the Diamond A Ranch
Paola: OK, so did they explain to you who pulled them over at gunpoint? Was it military?
Jose: It was the military and they had told me that they were looking for something, but Ed never did know that I had experienced that crash. He didn't know nothing about that crash. If he would have known, then he would have put more into that, we would have found out more about what happened. Maybe he could have kept that bar that he found. But he didn't know nothing until we started this, when you came in, you know.
Paola: I know, but you also told me something about you and your wife coming back in 1983 and you couldn't go there. You were going to show your wife where you were brought up. Did you also get stopped at gunpoint?

Chapter Twelve - A Trinity of Secrets

Jose: No, in 1983, that's when I went over to the manager from the BLM. That was when my boys were already grown, and I asked her if I could have a permit to get my boys to take a look at where I was born, you know.

Paola: In other words, you were never stopped? But did Reme ever tell you that he saw the file on this case? That the Governor of Washington State let him see the file, because he helped get her elected. So she pulled out a document, and *it's in the Atomic Energy Commission Files.*

OF COURSE! I thought. No wonder the case was never in Blue Book or any Air force records: the Atomic secrets were kept under their own system of P, Q, and R clearances. It wouldn't even be accessible to scientists with clearances above Top Secret from most Intelligence agencies, or the rest of the Executive Branch, all the way to the President, except by special requirement.

And that realization changed everything.

^ ^ ^

PART FOUR

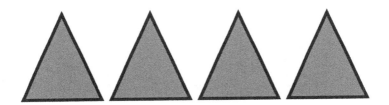

NINETEEN YEARS LATER

I am still of two minds as to whether or not this knowledge should be released into the world today. Once the step is taken there can be no return, and we have before us the horrible example of Atomic Energy.

Wilbert Smith
Head of the Canadian UFO investigation project, 1961

CHAPTER THIRTEEN

WRIGHT-PATTERSON AIR FORCE BASE

On the twenty-fourth of April in 1964, a Friday, I woke up at dawn in my motel room in Fairborn, Ohio, in time for an early breakfast with Dr. J. Allen Hynek and an 8 am briefing about UFO analyses at the Wright-Patterson Base of the US Air force. The two of us had left Chicago the day before, and we'd had a frugal dinner on the plane to Dayton, where Captain Quintanilla and his wife greeted us. This was going to be a short, but very exciting and momentous trip.

Late evening was spent at the Officers' Club of "Wright-Patt," relaxing and socializing with the men who ran Project Blue Book, the UFO office in charge of concentrating reports of unidentified flying objects from the military and the civilian population of the United States.

A short drive took us to another set of gates.

After the guards verified our clearances (for me, a simple two-day official pass, since I was not yet a US citizen) we were driven to the impressive, windowless aluminum building of FTD (the Foreign Technology Division of the USAF, not the florists…) which held the full, original records of Blue Book.

There were some 10,000 reports in the files at the time. Some of them consisted of a short teletype transmission; others filled 100 pages with maps, photographs and radiation readings. There is much mythology about those files, where they were, what they contained, who had access, and how they were processed.

Most Americans and many ufology buffs have been misled by serious-looking television personalities stating that the records, of course, were SECRET. If you follow any of the breathless video series about UFOs over the years, whether American or foreign, you "know" that the Blue Book files were super-classified, because the networks always write that in big red letters on the screen.

Even the latest *Project Blue Book* series on the History Channel perpetuates the myth.

They need to puff up their reporting and anyway, they know that's what the public expects.

Who wants to spend an hour watching a UFO story that's not promising some sort of great revelation? Who'd take time reviewing ordinary research data, no matter how fascinating?

The glory and mysterious aura of the promised "Disclosure" again...

Of course, if they had been secret, there would have been no way for me, a French citizen with no particular official credentials, to get anywhere close to that Air force Base, and especially to that building, since I didn't even have a US passport. In those days you had to reside in the US five years before you could even apply for citizenship. And mere citizenship wouldn't get you past the Marines with the machine guns at the gate.

If you don't understand the military, you may be excused for missing the fact that an unclassified, completely public archive may be housed inside one of the most highly secretive, most heavily defended buildings in the country. You needed a very good reason to be there, and a formal invitation.

The good reason, for me, was that my wife Janine and I had already spent years building, screening and editing an international database of UFO sightings and deriving statistics from it. As a result I had an up-to-date computer retrieval system for the global UFO Phenomenon, while the US Air force did not.

I was finishing a PhD in artificial intelligence and I planned to go on with increasingly comprehensive studies of the worldwide data, which

Chapter Thirteen - Wright-Patterson Air Force Base

already included some of the best cases in Blue Book after exclusion of the explainable ones such as meteors, planets, atmospheric phenomena and satellites, because I planned to redo the assessment work of their 10,000 cases, all of which would take time but was relatively straightforward to screen out with good software (and a horrible mess if you didn't have good software).

Many scientists in the US had qualifications equal to or higher than mine, and they could have requested access to the same files, but they were too busy pontificating on television, like Dr. Carl Sagan or *Aviation Week* journalist Philip Klass, explaining why there couldn't be any real UFOs in the files they did not (and would not) look at. The reason they didn't look at the data was that it would be a waste of their valuable time, which they spent "educating the public" by being on TV, which was a better choice.

Which, in the case of the brilliant Dr. Carl Sagan, was probably true.

The only scientists who went to "Wright-Patt" and looked seriously through the actual data after Hynek and me came much later. They were Dr. David Saunders of the University of Colorado and Dr. James McDonald, an atmospheric physicist at the University of Arizona. They were horrified.

At this point the curious reader is allowed to ask: "If the files of Blue Book were unclassified (with those few exceptions that involved detection by secret military systems such as radar prototypes), why were they located inside a windowless metal building on one of the largest, best equipped, most formidable Air force bases in the country? The answer is mostly one of convenience: UFO reports were gathered by local Air force bases and they were disseminated through a military communications network under precise military procedures and codes, so it was natural for the office (three officers and two secretaries, at the time) to be housed inside that building. But Allen and I had copies of the files at the observatory at Northwestern University, and with official permission we could have shared them with any *bona fides* request for access from an academic researcher.

This doesn't mean that approaching that office was as casual as a walk in the woods. Once you passed the gates, you went inside the FTD building where your credentials were scanned by another set of guards, and if you looked up at the ceiling you would see a jet fighter suspended above your head. Looking more closely you'd realize it wasn't any airplane you'd ever seen. It was a Soviet MiG with a red star on the tail.

That would put you in the mood, with a hint of what was going on inside that building.

Then you were briefed about the fact that conversations were prohibited outside the specific office to which you were assigned, and if you needed to pee, someone would take you to the appropriate place, stand next to you until you were done and bring you back.

Other than that, you could do what you wanted.

In that refined atmosphere, the Blue Book office itself was fairly relaxed for a group that was part of ATIC, the Air Technical Intelligence Center. On a day-to-day basis the business had to do with screening communications, either wires from the Air force network or letters from citizens. Many of the sightings they contained could be identified and answered fairly rapidly by reference to star charts, weather data or NASA schedules. Others would take a few phone calls and sometimes a site visit (especially when it had to do with a strange metallic residue, or actual traces in the ground, or photographs). And in the remaining cases, the staff would take an educated guess, which often proved wrong. **(32)**

When things escalated, the project would send out a request for Dr. Hynek to go to the place, contact local police or other authorities, and meet with the witnesses to compile a full report.

The office also contained a tall metal cabinet with double doors, its shelves lined with boxes of "stuff" people had found after supposed UFO landings, and shipped to WPAFB. Most of those were rocks with unusual shapes or colors that good citizens had found at the site and somehow associated with their sighting. Blue Book staff would identify the mineral (there were over a hundred PhDs at the Foreign Technology Division, covering every science, so Sergeant Moody or another officer just had to walk down the hall to get the answer) and a polite letter of thanks would be sent to the citizens who had reported the case.

Dr. Hynek and I had our own list of pending cases, but those were the tougher, unsolved ones for which we were eager to find out what the project had turned up in its latest investigations, so the hours went by quickly. Then it was time for another frugal lunch and Captain Quintanilla took us on a tour of the facility.

I was in for another lesson.

The weather was typical Ohio gray, humid, low clouds, lingering fog. We didn't stop anywhere close to the runways but Quintanilla pointed at

Chapter Thirteen - Wright-Patterson Air Force Base

the SAC (Strategic Air Command) Base through the car windows and the dampness.

I will always remember the rows of dark B-52s that looked like enormous night moths, their wingtips nearly touching the ground, next to strange little hills, bunkers inside which the crews were waiting for an event nobody wanted to think about, except that this was the Cold War and you couldn't stop thinking about those flights with no possible return, because ten minutes after you took off, Wright Patt itself would disintegrate in a gloriously luminous mushroom cloud.

A sentinel cradling a weapon was standing guard at the foot of every plane.

∧ ∧ ∧

We drove back to the office in a very serious mood and continued our discussions about the statistics the Project was reporting to Congress every year – usually a brief summary with a concluding statement such as "This year again, our Project has succeeded in accounting for 95% of all the cases submitted to the USAF by the American public and military services."

The implication was that, if a little more money was allocated, Blue Book would probably explain the remaining 5% by doing more detailed technical analysis, but the budget didn't make that feasible at the time.

Which is exactly what Congress wanted to hear.

If you can explain 95% of something, you should be able to extrapolate just a little bit and account for the whole thing, right?

Which is often true in science, and an acceptable short-cut even for juries in criminal law. Except that every course in Intelligence-101 teaches you exactly the opposite theorem: Your statistics don't work. Because the 95% were probably planted there by someone intent on fooling you or blinding you, and the 5% are designed to kill you. If you don't get that part, please read *A Bodyguard of Lies*, by historian Anthony Cave Brown: Adolph Hitler had 95% of the information about the Normandy landing because American Intelligence made sure he did.

You'll never forget that book, and you will also learn something about real computer history, more accurate than all the corporate promotional fluff on the web.

The Romans had said it best, and most concisely: *In cauda venenum*. All the poison is in the tail. That is still true. Courses in Intelligence also taught you that you couldn't do good analysis work and always please your bosses at the same time.

The officers assigned to Blue Book rotated through that office every two or three years, and they were very eager to go on to more exciting promotions in places like Japan, the European Command in Germany or (for the lucky ones) the Air force base in Honolulu.

Everybody knew Blue Book was a dead-end for any officer, but as long as they were there, the staff did want to do a good job, so Captain Quintanilla was genuinely curious about more scientific ways of processing his data, which was the main reason for our trip. We discussed sharing the software I was developing in Chicago and backing up Blue Book statistics with something more defensible, even if that meant that the number of "unidentified" cases would go up once the procedures were properly cleaned up; because there was always that faint possibility that "the whole dang UFO thing" was real, armed and dangerous.

Potentially, a political nightmare.

In which case there would be Hell to pay for the skeptics in Government and the debunkers in Academia who had denied the facts for years.

∧ ∧ ∧

The afternoon came to an end and Captain Quintanilla was kind enough to invite us to his home for dinner, music and some relaxation with his staff and their spouses, so there was dancing and good cheer. Everybody had such a good time that Hynek and I missed our flight, so it was decided we would take a later plane: more cocktails, and the ladies wanted to go on dancing, so we missed the 9:29 flight, and we ended up at a luxurious club for late drinks, all of which had the effect of ruining the excitement and genuine comradeship that exceptional day had created.

In the fake sophistication of that gold chandelier and red velvet club, listening to syrupy music as the hours turned, I felt I was finally understanding something about the Air force culture from the inside, a culture I still respected deeply: Neither the glorious, carefully crafted TV image of the brave pilots in their amazing machines, nor the dreadful misery of standing guard for endless hours at the foot of a bomber loaded with enough deadly technology to end all civilization, possibly by mistake

(Note), but something else: The conquest of what is best within all of us, the demonstration of intelligent mastery of our world, and the ability to peer over the horizon at mysteries like the phenomenon of unidentified flying objects, something the Air force actually understood (the pilots did talk among themselves, as pilots) but couldn't admit, when you looked carefully beyond the pabulum the Pentagon fed to the public.

Scientists didn't know what the pilots saw, and the public didn't either because television always filtered out the real thing, but Blue Book knew, deep down inside, that there was a real thing, even if it was a career-killer to say so.

Unbeknownst to us, as we sipped another Cognac in that fancy club, increasingly frustrated by the delays, the missed flights, the fake festive atmosphere, (and for me, the lingering image of those dark giant moths on that runway a few miles away, loaded with hydrogen bombs), something was happening in a dirty, dusty, empty lot south of Socorro, New Mexico, that was going to upset Blue Book and, beyond Blue Book, everything that was still holding together the fragile edifice of academic "explanations" of the UFO phenomenon.

∧ ∧ ∧

Note: The Strategic Air Command has been alerted in error three times since WW-2 with nuclear codes enabled, intent on bombing Soviet targets. In 1979, a repairman accidentally inserted a test tape into the computers at Cheyenne Mountain. By the time the mistake was found, fighter and bomber pilots were ready to take off with nuclear weapons.

The following year a faulty computer chip triggered a similar alert.

CHAPTER FOURTEEN

THE SOCORRO SYNCHRONICITY

The call didn't come in until the morning of Tuesday, April 28, 1964, four days after our visit to Dayton. Hynek and I had gone back to our daily work. I was watching graphic displays of an ongoing biological experiment for which I had written the program at the Faculty of Medicine of Northwestern University, a facility located downtown Chicago, when a black technician from the physiology department came to get me at the computing center: My wife Janine, who was at home since the recent birth of our son, relayed an urgent message from Mimi Hynek, who'd told her that Allen had just flown to New Mexico that morning after an emergency call from Blue Book, and that he wanted to talk to me.

When I managed to reach Mimi, she gave me the details as she knew them: Yes, an unidentified craft had landed there on a tripod gear, leaving deep marks; there were witnesses at the site including law enforcement officers, and Allen had been told to rush there by Captain Quintanilla in order to investigate. He'd taken the first available flight at O'Hare, he was now in Albuquerque, and he wanted to know if I could join him at Kirtland Air force Base that evening.

It was hard to reach Allen in the field, but when I did, things had already moved ahead. He gave me a summary: four days earlier, he said, on the evening of Friday, April 24th, a gray-white, egg-shaped craft with a strange red insignia had landed on four legs (not three), making deep imprints in the sandy, rocky soil of an empty lot with brush vegetation just south of Socorro, and setting the bushes on fire. A local cop named Lonnie Zamora had watched the whole thing and rushed to the site, thinking there might be an explosion at a nearby dynamite shack. Allen had interviewed the cop and believed him. The Air Force didn't.

Zamora had seen the object on the ground with two humanoid occupants of short stature standing next to it, but he didn't have much time to note details because the craft had taken off on a sort of thunderous vertical bluish beam that set the vegetation on fire, before flying away horizontally.

Three FBI agents were working in Socorro on a local criminal case at the time, so they rushed there to meet with Zamora and helped him document the traces.

Allen was interviewing many people; he'd already accumulated a lot of notes, and he'd soon be on the way back, so the best plan after all was for me to remain in Chicago and set time aside to build a framework for the complex testimony, including the field measurements and recordings he was bringing back. I assured him I would alert Bill Powers (the chief engineer at the Dearborn observatory) and our informal group of interested scientists at Northwestern and the University of Chicago to put together a background review and support a serious analysis as soon as he got back.

After hanging up I called together our small team; we started looking at maps and timetables. Some things jumped to our attention right away. Others only developed in our minds years later, after more extensive study and comparison with other cases, as in a slow process of integration. And other events simply baffled us.

One item struck me right away, as it did Allen: *If the craft had landed around 6:45 pm that Friday, then the event coincided with the time when Allen Hynek and I had just left the Blue Book office at Wright-Patterson Air Force Base near Dayton, Ohio.*

Our only joint visit to that office.

Was it some sort of test? A challenge to our research integrity? Or some sort of sophisticated joke, a twisted, yet serious riddle?

What a time to trigger what would become the best documented landing of an unidentified flying object!

Chapter Fourteen – The Socorro Synchronicity

In retrospect, there was another very important thing we all missed–including Captain Quintanilla, Project Blue Book, me and my computer retrieval software, Dr. Hynek himself and all the ufologists in the world; namely the now-obvious connection to the San Antonio Jose Padilla case of 1945, *because nobody in our group (or indeed, at Project Blue Book) would know about it for another sixty years.*

Hynek and Quintanilla, by then, would be dead.

Nobody knew about it for two good reasons: First, because what Jose and Reme had seen wasn't going to be reported anywhere by them until a new century had arrived, and second, because the very few people in the know at the Atomic Energy Commission and the still-secret Manhattan Project had no interest in talking about it to anyone. This state of affairs continued when the United States Air Force was created in September 1947.

The object was gone, the traces were gone, and the files themselves were gone until 2010 or so, when Paola Harris, who'd heard about it in Rome as an Italian journalist, began reconstructing the missing data and tracking down the witnesses who were still living.

She is the one who gave them a voice and went to the site to illustrate their experience.

The catalogue of Blue Book cases for 1945 only contains four reports. They would have come from those earlier Air force UFO projects, either *Sign* or *Grudge*, since Blue Book itself wasn't established until 1952. As a result, the way they were handled naturally varied with existing policies of the time.

From my own Air force database, those few reports were related to:
- a sighting in Habbebishopsheim (Germany) in April 1945: an American soldier reported a disk-shaped object that came down with an oscillating motion and landed, 35 km to the northwest of the town, a report apparently never followed up;
- a flyby case from Truk Atoll in the Pacific on May 2, evaluated as an ordinary aircraft;
- a civilian report from San Angelo, in West Texas, sent on July 17th, evaluated as a remote sighting of the first atomic mushroom cloud up in New Mexico, which is remarkable;
- and a report from a civilian in Westport, Indiana in September, evaluated as "imagination," a term that covered a wide variety of sins in the Air Force dictionary.

The Air Force had no UFO-related information at all from New Mexico for 1945, even in its classified files of *Sign* and *Grudge*. It wasn't able to make a connection between Socorro and the San Antonio case that had happened nineteen years before.

∧ ∧ ∧

The official story of the Socorro case is so well-known that we will only summarize it here from the best-documented sources. Officer Zamora, thirty-one year old at the time, gave his testimony multiple times (**33**) and his statements were always consistent with what he told his colleagues and the FBI when he was debriefed immediately after the sighting, late into the first night, and again Hynek in the following days.

The landing took place "in a ravine on the south side of that desert town, just before 6 pm on that Friday evening. Officer Zamora said he saw two small human-like figures outside the craft, but he subsequently has refused any further comment on that detail." (Ray Stanford, in his excellent book *Socorro Saucer in an Air force Pantry*, p.6, quoting an Albuquerque radio broadcast).

Much later, after he had resigned from the police force, Lonnie Zamora himself would describe the scene in a documentary for the show *Unsolved Mysteries* (Production 1532-8011) and explain that he was patrolling, and on his way to the Court House in his Socorro-2 squad car, when he saw a speeder come up the road, so he started chasing him, going up Park street. Suddenly seeing a big cloud of dust to his right, however, he thought he'd better go up there to check: a good police decision that he would later regret.

The dirt road was bad, and it took two or three trials for the 1964 Pontiac to climb up, kicking dirt and spinning tires, to an area from which he saw a white object to his left, which he first thought was an upturned car.

At that point Zamora called the dispatcher with a first report. But when he got up on top of the hill, he found himself looking down at a big white oval craft on the ground, unlike anything he knew. There were two short-stature "pilots" in white coveralls next to the craft.

They looked like small adults or tall children, three or four feet tall. They didn't wear any helmet, meaning that they could breathe our air. They seemed busy with something, and a bit surprised to see him.

Chapter Fourteen – The Socorro Synchronicity

Deciding to investigate, Zamora tried to call his dispatcher again to report that he was going to be out of his car, but somehow the radio had stopped working. So he left the patrol car, went towards the object and observed it for a short time until he heard a Boom! from the bottom of the craft, at which point he ran back in fear, diving and hiding behind his car in standard military practice, expecting the strange thing to explode. Peeking over his right hand from that position, he could see a flame coming out the bottom, as the object flew up to twenty or thirty feet in the air. It hovered there for a few moments, then took off slowly to the west, *in silence.*

Zamora's initial call to the dispatcher had alerted his colleagues. A police officer named Sam Chavez rushed to the scene: he found Lonnie very pale and in shock. He'd lost his glasses and his cap and looked like someone who'd been in a bad fight. There was smoke coming out of the grass in the arroyo where the object had landed, so the two cops walked down to the actual site with some misgivings, Chavez leading the way, fearing possible radioactivity or other unknown hazards. The mesquite was burning where the object had landed.

The two officers called for further assistance.

The call was heard by an FBI agent named Howard Burns who was at the police station about a Federal case. He decided to come over, followed by his colleague Mike Martinez, and later by a group of other people including an Army Captain named Richard T. Holder.

When interviewed by a TV journalist years later, Holder recalled how he'd become involved:

"We lived in Socorro at the time; we had two boys. We planned to go to Albuquerque the next morning, which would be Saturday morning, shopping. We're getting ready for bed early. About sometime between 7:30 and 8:30 that evening I got a phone call from my duty officer.

"I had two lieutenants there, and among the three of us we rotated the duty, which was basically an on-call type of affair. If they needed us, they called us at home or whatever phone number we gave 'em.

"And the lieutenant told me in a voice set somewhere between mystified and rasping, he didn't think I was going to believe it, but he had a call from an FBI agent that they had an unidentified flying object report, and he thought it belonged to the military. Since I was the senior military officer in the area—even as a very junior captain—I needed to come to the police station. So I called the police station, and the FBI agent said yes, that was so: please come down."

It was Holder who cordoned off the area that night. Assuming that some piece of test equipment had strayed off the range, he thought he needed some help and security. He'd arranged for the sergeant in charge and several military policemen to meet him at the site:

"We began to carefully take measurements, and collect samples of the debris and measure, let's say, what could possibly be called vehicular footprints in that area."

They were the first to pay attention to the indentations left by the object's legs, and the small footprints left by the occupants. Lonnie had seen an insignia on the side of the craft. Smartly, he grabbed the first piece of paper he could find and sketched it, afraid he might not remember it clearly later on. The insignia was red on white, about two feet by one.

He put it in his pocket without showing it to anyone.

Zamora went back to the Court House with his colleagues; they all went in, locked the door, and started bombarding him with lots of questions, Burns and Chavez leading the interrogation, which lasted beyond midnight. It must have been quite stressful, because in the absence of any logical explanation, some of the officers present went so far as to suggest that Zamora might have faked the whole thing in order to attract tourists to Socorro.

On one of the TV documentaries the interviewer asks: "Did you sleep well that night?" and Zamora flatly answers: "No, I didn't sleep at all."

∧ ∧ ∧

The next morning, Saturday April 25, 1964, is when Project Blue Book officially arrived in Socorro, in the person of Sergeant Moody and Major William Conners, a public information officer from Kirtland Air Force Base. Zamora had been called ahead by the Air Force office at Alamogordo, and told to be prepared to meet with them.

Captain Richard Holder, who identified himself as the up-range Commander at White Sands Stallion Site, joined the group along with FBI agent Arthur Byrnes Jr as they swept the area with a Geiger counter and found it at normal background levels.

Zamora was asked to go through his story again and in turn the Air Force told him about new evidence they had found; the site with the holes made by the landing gear had been roped off and measured, and *small footprints had been positively confirmed.*

Chapter Fourteen - The Socorro Synchronicity

Asked about dimensions, Lonnie compared the egg-shaped object to a VW bus, or perhaps twice the size of an ordinary car, possibly as much as 30 feet long by 10 feet high.

Which would place it in the range of the object nine-year-old Jose Padilla had seen and actually measured in San Antonio back in 1945: some 25 feet by about 13 feet high.

^ ^ ^

CHAPTER FIFTEEN

HYNEK'S FLIP

By Saturday May 2, 1964, Dr. Hynek was back from New Mexico. He spoke to me with some difficulty because he had dislodged his jawbone as he tripped over some stones at the site and had to find a local doctor in an emergency to get it wired. Overall, on the personal side, the trip to New Mexico and all the frustrating media interviews that followed had been quite an ordeal, as he related to us when we were able to get together.

Things hadn't gone well on the official side either; in fact, the case had become a hot source of embarrassment and restrained friction between Hynek and Quintanilla. Once broken, whatever mutual trust had existed between Project Blue Book, its scientific consultant, and our team in Chicago was never fully healed.

The bottom line was that Quintanilla was placed in an impossible position as head of Blue Book, and Hynek was in a comparably impossible situation as his scientific advisor. The Air Force was clearly frustrated with the events in Socorro because of the national dimension the story was acquiring in very visible media. The Army and the FBI were all over the case, local police was overwhelmed by the technical details, and the Air force had no progress to show on its own, bungled investigation.

I, the undersigned, do solemnly swear that on Friday, April 24, 1964, at around 5:50 to 6:00 P.M. there arrived at the Whiting Bros. service station (1409 N. California, Socorro, N. M.) of which I am manager, what I am positive was approximately a 1955 model greenish Cadillac, which was driven by a fairly tall man of medium build, sandy hair. His wife, and, I think, three children (all boys, the oldest of which appeared 11 or 12 years of age) were in the car with him.

The aforementioned man was quick to exclaim to me something like, "Your aircraft sure fly low around here!" He went on to explain, "Something travelling across the highway from east to west almost took the roof off our car...just south of town, north of the airport and right about in line with that junk yard. Looked like it might have been in trouble, since it came down west of the highway but not at the airport. Must have been in trouble 'cause I saw a police car head off the road and up a hill in that direction. Coming on into town I met another police car heading that way..."

When I suggested they might have seen a helicopter, he said, "That sure would be a funny-looking helicopter!"

My son, Jimmy also saw the aforementioned five persons, overheard the conversation, and thinks he remembers their car had a Colorado license plate.

Jimmy and I did not think too much about the man's report, because we had not yet heard of Lonnie Zamora's report. As soon as we heard what Lonnie had seen, we knew the tourists who had stopped by our station must have seen the object Zamora saw heading for a landing in the ravine.

That is why I reported the incident with the tourists-- because I wanted people to know that other persons saw the same object Lonnie Zamora saw. Lonnie wasn't just 'seeing things'.

This affidavit shall be considered as public notice of the facts as set down herein and of my personal testimony to them.

Hence, duly sworn, in the presence of the witnesses listed below, on this eighth day of January, 1967, I make this testimony by signing hereon:

Opal Smith
Opal Grinder, Soccoro, N.M.

Witnesses:
Robert McGarey

Fig. 25. The Opal Grinder Affidavit.

Chapter Fifteen - Hynek's Flip

Project Blue Book had called reinforcements in the person of Moody and Conners, but we were told they had done nothing to improve the situation. Dr. Hynek, who had not even been alerted to the case when it first happened, had been sent there as a last resort.

It may be that, given the unexpectedly high media visibility of the case, Quintanilla was unhappy with the shoddy reports he had been getting up to that point. He still hoped that Dr. Hynek could be counted on to bring a rational aura to the investigation and to provide a nuanced scientific interpretation of the sighting and the traces.

Was Dr. Hynek also setup to serve as the fall guy if the conclusions of Blue Book turned out to be wrong?

Things didn't quite work out that way.

Dr. Hynek had been met at the Albuquerque airport by Major Conners driving an official car and things went wrong right away. To begin with, they had a flat tire on the way south and there was no spare in the car, so the Air Force's top scientific consultant on the case had to hitch a ride from a passing motorist to get to Socorro, where he finally met with Zamora and Chavez. By then, both officers were very upset at the hostile way they had been interrogated by Moody and his Air Force colleagues. There was tension and resentment all around.

During an interview with the TV show *Unsolved Mysteries*, years later, UFO historian Jerome Clark would put it more directly and, in my own recollection, more accurately: "It was sort of symbolic of the general incompetence of the Blue Book investigation that they couldn't even transport their leading scientific consultant from one spot to another. By the time Hynek got to Socorro and met with Zamora and Chavez, they were extremely annoyed at the way the Air Force had treated them."

Hynek took control of the situation by sending Moody away and driving straight to the site of the landing with Zamora and Chavez.

Since several days of confusion had already elapsed, however, it was very difficult to conduct a true physical examination there, because neither Moody nor the local law enforcement people had been careful about preserving the area that the FBI had properly roped off that first evening. It had been trampled by onlookers and souvenir collectors: little of the initial evidence was left.

That wasn't Hynek's only problem, however. Reports were circulating that other witnesses had seen the object on its way to the site. One of them came from Opal Grinder, the manager of a service station on North

California Street: just before 6pm that day, a man driving a green Cadillac with his wife and three boys had stopped and complained to him about "something travelling across the highway from east to west" that "almost took off the roof of our car."

The man wanted to report it, thinking it might be an aircraft about to crash, and Grinder eventually filed an official affidavit. (Fig. 25) The Air force ignored it carefully.

That was only the beginning. Captain Holder's official report mentions that "Upon arrival at the office location in the Socorro County Building, we were informed by Nep Lopez, sheriff's office radio operator, that approximately three reports had been called in by telephone of a blue flame or light in the area (of Zamora's observation)... the dispatcher indicated that the times were roughly similar."

Those three reports had been made before the incident was publicized. They were never seriously taken into consideration.

∧ ∧ ∧

A map of the landing site published by independent researcher Ray Stanford shows the five Colorado tourists driving north on Highway 85, but it also shows two other witnesses on Highway 60, namely 24-year-old Paul Kies and 26-year-old Larry Kratzer, both from Dubuque, Iowa, who observed the landing itself from the southwest.

Ray Stanford's careful research turned up the fact that they were headed north-northwest on highway 60 into Socorro (see map, Fig. 26). They said the object was accompanied by flame and brownish dust, possibly kicked up by Zamora's car.

Overall, Mr. Stanford identified eleven visual witnesses and two auditory witnesses in addition to Lonnie Zamora himself.

When Hynek proposed to extend the investigation to these other witnesses, a pretty obvious step for any scientist, the Air Force flatly denied the authorization, so their top scientific consultant had to rely on second-hand data, including the good independent research done by Ray Stanford, who had arrived in Socorro at about the same time. Stanford, as we will see, took a parallel approach but it was wider and, frankly, more scientifically based.

The Air Force also forbade Dr. Hynek to look at comparable sightings, such as historical observations of oval objects with strange blue "flames" in our available databases.

Here again, by presenting the case as a very brief, one-witness observation by an ordinary Latino cop, Project Blue Book could reassure Congress and the august, skeptical American academic community about the lack of any serious threat either to the Nation as a whole, or to cherished scientific theories. The fact that the pattern matched no flying device known, then or now, did not bother the Foreign Technology Division: It was too far outside its normal purview. (Note)

An unintended outcome of that flawed strategy was my newfound ability to finally convince Dr. Hynek of the similarity of American landings and European cases—such as those recorded by the dozen in the French UFO wave of 1954 that I kept presenting to him. Those were some of things we had just discussed at Wright-Patt, where the staff had practically no background knowledge of UFO observations from other countries.

It also led him to the realization that Project Blue Book, while it retained some value as a channel for the monitoring of reports across the country, had become scientifically irrelevant.

And Hynek saw that he was being used to prop up inaccurate theories about the observations that might keep the academic community smug and unconcerned, but violated his own standards and his early motivations for getting into a scientific career as a young astronomer in the first place.

As we saw above, Ray Stanford, who did the most thorough independent analysis of the Socorro case (**33**), identified a number of independent witnesses. Their testimony was critical in convincing Hynek, who might not have taken a stand on Zamora's report without it: He often retreated behind another Latin legal principle: *Testis unus, testis nullus*: A single witness is no witness... which was only true if you studiously avoided looking for the other people.

And in fact, Zamora had been the only witness, when it came to the dwarfish pilots of the white oval craft, if you wanted to pick at straws.

Mr. Stanford also conducted his own analysis of the physical characteristics of the object itself. In his excellent book, he quotes military specialists who examined the traces and computed that the weight would

---

Note: It is interesting to note that the 1967-69 Colorado study under Professor Condon, which claimed to review the American data, and was applauded by the Academy of Sciences and the *New York Times*, never touched the massive records of the Socorro case and didn't even include it in its index.

need to be in the range of 8 to 10 tons (16,000 to 20,000 pounds) in order to produce the four prints measured by the officers at the site.

He also commented on the compatibility between the craft's apparent density and its shape and volume as accurately described by Lonnie Zamora.

The object was denser than ordinary aircraft but the values fully supported the consistency of the parameters in Zamora's report.

∧ ∧ ∧

The final, official report entered into the Air Force files by the Blue Book staff contains gross inaccuracies, and it is hard to call them unintentional.

Ray Stanford lists them in his book:

- The report states that Zamora was driving north in pursuit of a speeder. Actually, he was driving south, as one look at the map will confirm.
- Even the road is wrong: Zamora was not on Highway 85 as stated, but on Park Street. He had said so himself in the course of the very first interrogation, *on the record.*
- The report mentions that he saw "flames," while he only spoke of "a blue flame."
- The Air Force writes he was 800 feet away from the object when he first saw it. All police indications agree with Ray Stanford about a distance of only 450 feet or so.
- The report goes on with the statement that Zamora saw "one or two" figures "in coveralls." That's not what the witness said. He stated he saw two figures in "what *resembled* coveralls." He added the clothing was white, which the report omits.
- Zamora never said he saw "smoke" coming out of the bottom of the object.
- The fact that the occupants were dwarves is never mentioned anywhere.
- The report asserts that he parked his car 150 feet from the gully, which means he would have observed the craft from about 200 feet. In reality he parked on the very edge of the gulley, placing him only 50 feet away, an error by a factor of four!

Chapter Fifteen - Hynek's Flip

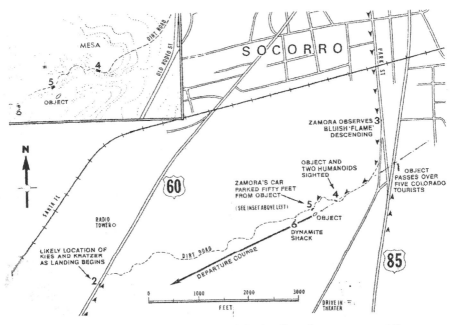

Fig. 26: Ray Stanford's reconstruction of the landing site geometry at Socorro.

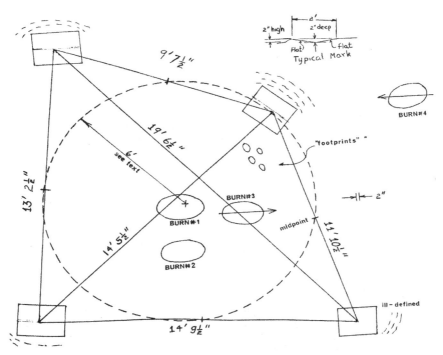

Fig. 27: Bill Powers' analysis of the remarkable traces left at Socorro.

Ray Stanford confirmed the distances involved in a message to me: "Hynek asked me to begin pacing where Lonnie's feet had been, and he watched me pace off the distance from there to where the 'exhaust' had met the ground. Allen and I agreed, based on my paces, that Lonnie had been about 36 feet from the center of the object's above-ground location. Lonnie agreed on that distance figure."

- The Air Force writes that "there was no evidence of markings of any sort, other than 'shallow' depressions at the site." Mr. Stanford asks, "What about the vitrified sand and rock?" What about the large bush that was sliced in half by the blue-white flame?
- The official statement that "No other witnesses to the object reported by Mr. Zamora could be located" is a blatant lie. The Air Force knew of the other eleven witnesses.
- The inconsistencies in the edited or redacted statements point to a desire to minimize the strangeness and to impugn the reliability of the case; combined with the censoring of the observations by the other witnesses, this amounts to nothing less than blatant disinformation, directed at the US public and the scientific community.

These accumulated errors would make it practically impossible for any independent scientist to review the facts of the case and assess the conclusions of the Air Force report. In fact, taken together, they would give the academic community the impression of a single-witness, borderline observation of some test vehicle from one of the classified military labs next door.

Which may have been the intended effect.

In a book summing up his experience as the long-time scientific consultant to Project Blue Book, published in 1972 (**34**), Dr. Hynek finally felt free to voice his personal concern, as a researcher, about the conditions thrust upon him by the military.

In an official letter to Air Force Colonel Raymond S. Sleeper, dated October 7, 1968, he wrote:

"I sincerely hope that at long last... I may help transform Blue Book into what the public and the scientific world have been told it is...an investigative organization dedicated to the defense of the country but doing a good scientific job also...It is time that Blue Book no longer be called, as some wag has done, *the Society for the Explanation of the Uninvestigated.*"

Chapter Fifteen - Hynek's Flip

Returning to the Socorro case in the same book (pp.144-145) he observed:

"I tried my best at the time to induce the Air Force to make an Intelligence problem of finding the missing witness, but they evinced no interest whatsoever. At the time I thought that, had this been a Federal case involving narcotics or counterfeiting, the FBI would certainly have located the missing witness. Because it was merely a UFO case, the usual pattern of doing nothing was followed."

^ ^ ^

Back in Chicago, we recompiled the elements we had, as Dr. Hynek returned from New Mexico, still angry and encumbered by his painful, healing jaw. The discussion set aside the Air Force concern with superficial explanations and public relations. Instead, it branched off in two directions: First, a step-by-step re-examination of everything Lonnie Zamora had reported, including his personal impressions and feelings before, during and after his observation; and second, an attempt to place the object, as a piece of technology, in the context of what physics and aerodynamics could tell us about such a device.

Both of these pointed to the honesty and reliability of the main witness, but they added another layer of engineering significance that the Air Force had completely missed.

At the site, Dr. Hynek had toyed with the idea that the craft might be a test article for a lunar landing, or an "Air Force Pogo," a prototype for a craft with a central cylinder that would propel it upward in the same way as a Pogo stick. The citizens of Socorro, located so close to the immense facilities of the White Sands test sites, were used to overflights from all kinds of airplanes, helicopters and other devices undergoing testing, but those objects had no reason to land in a vacant lot in town, and especially in a zone that wasn't even level. It seemed whoever piloted that object trusted his landing gear and knew that he could rely on its sophistication to such a point that he didn't even care about the roughness of the landscape under his craft.

The files in fact contain an interesting interview with a police officer named Ted Jordan, who was also on patrol that evening. He'd heard Zamora's excited call for help on the radio, and joined him on top of the gulley. He had carefully examined the elements present around the vacant lot, treating it as carefully as he would have treated a crime scene: A meticulous policeman, he noted every piece of cardboard, and the odd newspaper. He was struck by the fact that none of them were burned, but they had obvious scorch marks; and they hadn't been moved from their initial location when the craft took off, which he could tell by inspecting the appearance of the soil protected from the elements directly under the cardboard or other object.

He testified:

"I pointed this out to, ah – there was, ah, an investigator down here named doctor Allen Hynek; and, ah, he had advised me that what it was, was an Air Force Pogo: a Vertical Take-off and Landing jet had landed there. And we disagreed on that point. And I told him that I was no Air Force expert, but I would think that anything with enough thrust to lift an aircraft of that size would certainly have blown away a piece of cardboard. So I, ah, I believed that whatever it was didn't have a propulsion system, such as a jet engine. I don't know what in the world it was, but Lonnie had told me that, after the thing started up, it had no sound. It, ah, the sound quit on it. When it departed."

The idea that the craft seen by Zamora could be an Air Force Pogo was reviewed by our little group, but it didn't survive Bill Powers' engineering analysis: Both the soil traces and the absence of sound in horizontal flight also negated the concept of a rocket-propelled craft, as officer Jordan had said in his site analysis report, remarkably void of any preconceived notions.

The shape of the craft itself, with its aerodynamic envelope, bore no relationship to the open, tinker-toy structure of the lunar lander that would take Armstrong and Aldrin to the moon in July 1969, five years later; with a fixed landing gear that made the mission extremely dangerous.

Figures 26 and 27 summarize the physical parameters of the case as we reviewed them. The landing took place, for no logical reason we could find from the behavior of the supposed occupants or pilots, in an uneven vacant lot I've had the pleasure of visiting on the outskirts of Socorro, at longitude -106.900 (106°53'53"West) and latitude 34.038 (34°02'24"North).

Mr. William T. Powers, the chief engineer for Dearborn Observatory, carefully re-assembled everything known about the ground traces (Fig. 27) and made a most interesting discovery: Although the extension of all four legs varied, given the rough terrain, the diagonals of the quadrilateral figure they formed intersected *exactly* at a right angle. This was another technical feature that could not have been anticipated by Lonnie Zamora or anyone else involved. But the implications in terms of the sophistication of the craft were interesting.

There's a theorem in geometry that states "When the diagonals of a quadrilateral are perpendicular, the midpoints of its sides and the feet of the perpendiculars dropped from them on the opposite sides all lie on a circle described about the mean center of the vortices."

In the case of Socorro, it meant that the figure carefully drawn at the site by the police was a very special case: on their map the center of the circle in question was found to be directly over burn no.1, corresponding to the center of gravity of the landing object.

Bill concluded (**35**, p.50): "This random-looking placement of the landing pads would result in equal distribution of the weight of whatever those pads were supporting."

He added, comparing the Zamora object to a space probe:

"The object making the marks either supported a large weight or hit very hard, since the soil is dense. NASA has concluded that the Surveyor pads sank about two inches into lunar soil with a bearing strength of five pounds per square inch; the gravity (on the moon) is six times lower, but the pads are only about one-fourth of the area of the marks at Socorro; we must assume that the force was equivalent to gentle settling of at least a ton on each mark."

So much for the Air Force "Pogo stick" idea.

Whatever landed at Socorro was more sophisticated than anything the US sent to the Moon before or after 1964. It did not use rocket propulsion, but a type of particle or energy beam of remarkable characteristics. It stopped making any noise once it was airborne; and it didn't appear to have any concern in preserving its stability when it landed in any type of terrain.

As my old friend Bill Powers aptly observed, as an admiring tribute from one seasoned engineer to another, "We must conclude that whoever designed that landing gear must be an interesting fellow."

∧ ∧ ∧

Some of the remaining puzzles about the Socorro landing might be solved if we had more extensive evidence of the marks, including footprints and burn marks on plants, in the form of detailed photographs of the site. That is the kind of hard data that scientists reviewing the case would need to have to make an independent judgment.

As it turned out, there was such evidence at one time, thanks to New Mexico State Trooper Ted Jordan, the man who, apart from Ray Stanford, did the most extensive examination of the area as an officer trained in crime scene documentation. So, why don't we have that stack of photographs in the file?

Jordan was asked that question in a later interview and replied:

"I had a roll of, ah, thirty-five millimeter slide film in my, ah, camera that I had taken pictures at the site, this aircraft landing, I had taken most of them. I usually carry a thirty-six exposure roll in there. And I'd taken most of them at the site that day; there was a lot to photograph; and I took lots of pictures."

"And then the—the Air Force ended up with them. And I don't recall, right now, whether, ah, they were given to Captain Holder or one of the Air Force officers. I do not recall that. But I do recall that they told me that they would, ah, develop them and send me copies of them, ah, it never occurred."

"And when I asked, later, about it, they told me that the film was no good—ah, it had been ruined by radiation."

"And I never got any, ah, film. I never even got—I never received any explanation—except verbal—that they were ruined by radiation, and that I wouldn't be receiving any, ah, film. When I asked them to send me the, ah, 'go ahead and print them and send me the blanks', they wouldn't even do that."

The gentle reader will recall that the presence of radiation had been immediately, and very professionally tested on the first day—not an unusual or difficult thing to do when you live less than an hour away from the world's first Atomic Test Site—and there was no radioactivity. Officials were simply lying all the way down the line.

Best of luck to the innocent scientists who would like to conduct an independent review of the most significant unexplained case of a UFO landing, complete with occupants and physical evidence, in the files of Project Blue Book.

∧ ∧ ∧

Chapter Fifteen - Hynek's Flip

We cannot leave the picturesque environment and the aerospace mysteries of Socorro without mentioning one more fanciful "explanation" for the Zamora sighting mentioned, this time, in a supplement to the local Socorro newspaper, *El Defensor Chieftain* for November 15, 2003 under the pen of James Easton, a freelance journalist who lives in Scotland.

Mr. Easton writes for a British magazine called *Fortean Times*, supposedly dedicated to continuing the work of celebrated researcher Charles Fort. An extraordinary compiler, Fort spent much of his life assembling careful records of anomalies, such as unexplained flying objects in the nineteenth century, as well as curious biological, medical or atmospheric effects described in the scientific press, which defied contemporary explanations. Many of them still do, such as observations of ball lightning, and unusual meteors. These old records are precious for today's professional physicists.

In recent times, however, *Fortean Times* has increasingly applied itself to the more rewarding chase of bizarre or titillating anecdotes that it could hold up as an example of an anomaly in the eyes of the naïve reader, but for which reassuring explanations were found by the clever, well-informed skeptic. While this is certainly useful in quelling idle speculation and rumors, and while it makes for fun entertainment, it tends to be the opposite of what Charles Fort tried to achieve all his life: He saw himself as a censor but also as a servant of science, generally starting from unsolved, neglected but authentic anomalies with real observers, reviewed and published in the scientific press, such as dozens of observations of unexplained luminous objects tracked (but too often ignored or superficially brushed off) by professional astronomers, and he urged that they be studied.

Niels Bohr, decades later, would support that approach when he taught that "there can be no science without an anomaly."

Fortean Times published Mr. Easton's statement that he had finally determined that what Lonnie Zamora saw was nothing but a balloon, a fact that "evidence, witnesses, CIA responses make clear": "The Socorro 'flying saucer' was a hot-air balloon." Problem solved!

He bases this conclusion on the statement Zamora made to Ray Stanford about the slow but noisy departure of the object with a flame that didn't exert any thrust upon the ground, followed by silence as it hovered and flew away horizontally. Suppose, he says, that the flame came from a fire inside the egg-shaped craft, raising the air temperature inside (without setting the balloon on fire or asphyxiating the two pilots?) with the exhaust somehow deflected down: *Voilà!* The balloon would rise majestically over the heads of the befuddled cop, then it would move away in silence.

Hot-air balloons, however, are designed with a nacelle where the occupants can breathe normally while the envelope above is filled with hot air, a very different configuration than an enclosed, rigid ovoid. Also, hot-air balloons don't weight nine tons, and they don't land on heavy, retractable metal legs of high geometric sophistication.

Seeking confirmation for his thesis, Mr. Easton had contacted Ed Yost, well-known New Mexico balloonist who piloted the first hot-air balloon flight on October 10, 1960, in a prototype from Raven Industries. He could not recollect any of them landing near Socorro, however. Next, Mr. Easton tracked down a man named Jim Winker at Aerostar, which had taken over Raven. Mr. Winker confirmed that advanced hot-air balloons became popular in the early 1960s and some of them were whitish like the Zamora object but—importantly--"they had a platform for the crew." Which sounds like careful support for the reality of hot-air balloons in the area of Albuquerque, but Mr. Easton himself notes there were only about 20 of these new models in the entire United States. (The first-ever crossing of the English Channel by a hot-air balloon had taken place in April 1963).

As for the CIA, *it was very careful not to confirm anything at all*, contrary to the screaming headline in the Socorro paper that implied government support for the Easton thesis ("CIA responses make answer clear"). On the contrary, the Agency had soberly stated that it "couldn't confirm or deny the existence of records responsive to the request." A second set of questions simply received the same treatment: the CIA stressed again that its earlier response still applied,

Which makes sense simply because the CIA has no direct interest or jurisdiction in domestic matters, including hot air balloons...

So who is twisting the truth here? The journalist, or the CIA?

The reader should note that I am not excluding the possibility that advanced rigid balloons might have been tested around the White Sands facility in 1965: That hypothesis would have made some sense at the time, except for the fact that the local police officers, the Army personnel from the base, and Dr. Hynek, all were familiar with balloons. If such advanced balloons were being tested at the time, what was their mode of propulsion? Certainly not "hot air." And why are they not in operation today, 56 years later, if they work so well?

Chapter Fifteen - Hynek's Flip

By the time detailed inspection of the site had been conducted, particularly by State Trooper Ted Jordan in his highly professional, meticulous inspection of the environment, multiple pieces of evidence negated the balloon idea but they only came out later, and researchers like Mr. Easton may be forgiven since they did not have convenient access to them.

Interviewed for the show *Unsolved Mysteries* some years later, Jerome Clark, an associate of Dr. Hynek with vast historical knowledge of the field, pointed out that "the Air Force went to considerable effort to find an aircraft that looked like the object that Zamora reported. It found that a lunar landing module was being tested at a base in southern California, but there was no evidence whatever that any such craft was being tested in the desert of New Mexico on April 24, 1964, or anywhere close to it. The Air Force was also intrigued by the strange insignia that Zamora had reported on the side of the object, and went through great efforts to find something comparable somewhere in aviation technology. It couldn't find anything. The insignia remains one of the many interesting puzzles about the Zamora case. Nothing like it actually has been reported since. And nothing like it had been reported before."

In addition to the data I have quoted, the reader will recall that the official analysis showed that the object weighed 16,000 to 20,000 pounds

Fig. 28: (A) Dr. Hynek's own record of the Socorro insignia, (B), the currently accepted "official" insignia, and (C) the initial *false* design BlueBook gave to the Press in 1964.

(8 to 10 tons). In separate calculations, Ray Stanford, in the very book quoted by Mr. Easton, had published an estimate for the pressure on each of the feet corresponding to a weight of one ton.

It would take a lot of hot air to lift such a craft, a lot more than it does to float another reassuring UFO "final explanation" in the local paper, or in a skeptics' magazine far away from the actual scene, on another continent.

∧ ∧ ∧

The real problem for Mr. Easton's hypothesis, *and all the others*, comes in the form of a totally independent testimony given, years later, by a top expert from White Sands, Army Major Richard Holder, retired. I have already mentioned his name in connection with the Air Force on-site investigation. At the time, in 1964, he was a very junior captain, assigned as the up-range commander at Stallion Range Center, the northern extension of White Sands Missile Range.

He humorously described his domain as "roughly two hundred thousand acres of rattlesnakes, sagebrush, beautiful country, lava rock." He had about 120 military personnel under his orders, responsible for logistic support and recovery of the up-range command.

On the Monday following Zamora's sighting, which would have been April 27th, Captain Holder got a direct report of an unpublicized incident that occurred on the range, on a vector that the Socorro object had to follow if it continued in the direction Zamora reported.

The witness was a very senior, non-commissioned officer, close to retirement, very experienced, and a much-decorated combat veteran. *His position was that of a master mechanic.* He had initiated a policy, supported by Holder, of checking every vehicle that went downrange to be pronounced as secure and to have a radio installed. Upon departure, they checked with the military police at Stallion Point, checked again at mid-range and checked in when they arrived at White Sands, all for safety purposes.

Nobody wanted a crew or a family stranded in the desert: it is important to remember that the White Sands Missile Range, the largest military installation in the United States, covers 3,200 square miles, bigger than the State of Rhode Island and the State of Delaware, *combined.*

The evening of Zamora's sighting, the officer was driving on a range road, about halfway between the mid-range point and Stallion Range

Chapter Fifteen - Hynek's Flip

Center, about 37 miles South of that point, when he saw a light to his right (West) up against the mountains, gradually increasing in intensity. He described the colors of the light very much as Zamora had done. As the source became intense, the military vehicle stopped.

Everything electrical aboard ceased to function: the radio, the headlights and the ignition.

The witness got out of the vehicle to watch the light. As its intensity diminished, the headlights came back and the radio was heard again, with static. The officer started the car on the first try, drove it down-range, put it in the shop, and had it totally examined: There was absolutely nothing wrong with the vehicle. The incident took place about an hour after the Zamora observation. I can find no trace that it was ever reported to Dr. Hynek, or to Project Blue Book.

∧ ∧ ∧

So we now have "missing" evidence in the form of the photographs of the traces and the environment taken by a State Police officer, made to disappear by Project Blue Book, with a lame explanation (radioactive fogging) that conflicted with their own measurements.

We also have evidence of the fact that Dr. Hynek was never permitted to mention the supporting witnesses he knew about, or to go in search of others whom he didn't yet know about, even though his mission was supposedly to perform a scientific investigation.

None of this is hearsay, but direct testimony by the principals, *on the record*.

And third, we have evidence that a highly-competent source, a senior Army officer, a combat veteran specialized in military vehicle engineering, experienced the total failure of the car he was driving on the White Sands range while a light attached to a flying object increased in intensity and illuminated his position, an hour after the Lonnie Zamora sighting *and on the trajectory reported by the policeman*.

There is a fourth series of events that relate more directly to the scientific problem itself, that of material evidence. Since it touches directly on the work my colleagues and myself are currently doing, I tend to place special attention on it.

The case derives from the efforts of Mr. Ray Stanford, whom we've quoted before, to conduct independent research on the object seen by officer Zamora. On the Wednesday following the sighting, the 29th of April, he had gone back to the site to check up on a statement Zamora had made, about a particular rock partially broken by one of the craft's landing pads as it took off.

Zamora's insight was prescient: When he examined the broken surface carefully, Stanford found that particles of a bright metal were attached to it, evidently the result of hard mechanical scraping. Since the area had been left open to the public and not restricted in any way, he simply took the rock and put it in his car, hoping for an opportunity to conduct a close examination later. When he was able to do this, a few days later in Phoenix, he verified there were "pseudo-metallic materials" clinking to the stone.

On July 31, 1964, according to his book, Ray Stanford met with his NICAP (**36**) colleagues Messrs. Richard Hall, Walter Webb and Robert McGarey in Washington, D.C. and they drove out to the NASA Goddard Space Flight Center where materials expert Dr. Henry Frankel had agreed to meet with them to discuss the possible analysis of the residue.

Stanford's account of the meeting is fascinating, because it makes it obvious that the scientists in that NASA unit not only took UFO data seriously, but were evidently well briefed about the Socorro event.

Looking at the rock under a microscope, the metal particles were obvious, and Dr. Frankel immediately noted that the material must have been in a molten state when scraped onto the rock, after which he suggested that the visitors go to lunch, where he would meet them once the material was collected. He also agreed that Stanford could retain half of it.

Frankel never joined them at the Goddard cafeteria.

When they came back to the now-empty lab, they found that he'd left a single technician in charge. When the man returned the rock to Stanford *it had been scraped clean*, contrary to the oral agreement the NICAP team believed they had reached with NASA.

Dr. Frankel eventually returned, according to Stanford's book, and he said they had to take all the material "to make an accurate analysis." The sample was going to be placed under radiation that same afternoon, where it would remain the entire weekend. On Monday, it would be removed for X-ray diffraction tests that should tell the elements it contains. He added: "I will notify Mr. Hall of the results by telephone and if you, Mr. Stanford, will call me on Wednesday, I should be able to tell you something very definite."

Chapter Fifteen - Hynek's Flip

Ray Stanford was puzzled, but the two ufologists, Richard Hall and Walter Webb, advised to just trust NASA because the contact with Dr. Frankel was important to the NICAP organization.

Indeed, on Wednesday, August 5, 1964, Stanford did call Frankel, who immediately said, "I have some news that I think will make you happy." And he went on to explain that the particles that were clinging to the rock's surface contained zinc and iron, but *the ratios did not correspond to any material known to be manufactured on earth*, at least per the charts at Goddard, which were the most complete in the world. **(37)**

Dr. Frankel went on to say that the alloy would make "an excellent, highly malleable, and corrosive resistant coating for a spacecraft landing gear."

He added the lab was still working on the trace elements that were present in small percentages, and would have an answer in a week. But when Mr. Stanford called on August 12th, the department secretary told him her boss was unavailable, and she asked him to please call once more, the next day.

On Wednesday, Stanford was told to call again at the end of the following week.

He negotiated a Monday phone appointment, but when he called from a phone booth in Texas on that day as he was driving back to his home in Arizona, the secretary told him that "Dr. Frankel is unprepared, at this time, to discuss the information you are calling about."

Could Ray please call back the next day?

Ray Stanford must be a very patient man, because he followed those instructions, from Corpus Christi this time. He was then told that Dr. Frankel was "in a top-security conference."

So Stanford suggested perhaps the scientist could call him instead, at his Arizona number, whenever he had the information and was available to discuss it.

Frankel never did. It seems that, at that point, responsibility for the sample had been removed from NASA-Goddard and taken over by an agency with the "Need to Know" about UFO materials. **(38)** Days passed, and eventually an official-sounding man named Thomas P. Sciacca, Jr. called to tell Stanford that Frankel "was no longer involved in the analysis."

Not only that, he said, but "everything you were told earlier by NASA was a mistake. The sample was determined to be silica, $SiO2$." **(39)**

According to Mr. Stanford, what had started as an informal agreement to do some testing on a strange scraping as a favor to some researcher friends had been escalated, first, into a full-scale official NASA analysis, and then escalated again to a level where it was felt necessary to provide a cover story not only to the confiscation of the evidence from Goddard, but to its fundamental nature.

As often happened in such situations, nobody cared about the fact that the official "final answer" about the composition of the sample was plain stupid: They knew everyone would be happy to see the problem go away, even at the price of a breach of professional ethics, followed by the publication of scientific nonsense.

Any NASA chemist who was so confused as to analyze a mundane sample of ordinary silica and conclude it was composed of zinc and iron, plus those undisclosed "trace elements," should have been reassigned to the maintenance of the Xerox machines.

Yet "government officials" stuck to the story, shutting up pesky journalists.

They were even quite prepared to put it in writing, borrowing NASA stationery for the occasion. Interestingly, when NICAP published its final report on the Socorro case in the *UFO Investigator*, it only quoted the "silica" cover story, which it must have known from detailed, first-hand information to be untrue. (*UFO Investigator*, Sept-Oct. 1964 issue).

With the benefit of over half a century, it appears that Dr. Frankel and his NASA colleagues may have been sincere in their interest in the Socorro metallic evidence, but they were caught in demands from an unspecified higher authority.

The situation seems to change a bit today, as some highly credible admissions by government insiders have been allowed and a few files opened, along with interesting videos of unexplained objects posted on the Internet. But the academic community has yet to take notice.

∧ ∧ ∧

CHAPTER SIXTEEN

GLOBAL PATTERNS AND A SURPRISE THIRD WITNESS

As an information scientist, I have been trained to look for patterns—in astronomy, in medicine, in business—rather than extrapolating from a single observation, no matter how interesting. In this respect, I think we've made two major mistakes in looking at the UFO phenomenon. The *first* major mistake is to assume that the phenomenon, if real, *must* consist in space-travelling Aliens from another planet—to the exclusion of any other possibility; and the *second* major mistake has been to wait for the one, single, absolutely perfect case you can hype on TV and take to Congress, that will "instantly" prove the existence of UFOs, convince the skeptics, lead scientists to file patents for space propulsion or new weapons, and unveil the final key to the mystery.

Unique discoveries based on a single observation or object do happen, but they are rare: the discovery of the Rosetta Stone that made it possible to decode the hieroglyphs, the sudden fall of dozens of hot mineral fragments bombarding the small French town of *L'Aigle* that damaged many houses in a few minutes and proved the existence of meteorites, or the accidental placement of a piece of radium next to unexposed photographic plates in a laboratory drawer, that fogged up the negatives and made obvious the phenomenon of radioactivity, were examples of lucky mistakes or fortunate accidents that upset established Academic catechism and started new disciplines. But they only made sense because there were pre-existing patterns of many observations that needed to be explained, and a pre-existing theoretical scaffolding that could be adapted to handle them.

For centuries, academic astronomers had denied that stones could fall from the sky, blaming the reports on the supposed stupidity of farmers who kept picking up strange stones: everything seemed to tell them, in everyday life, that there were no stones in the sky. But they should also have known that, if farmers hadn't been just as smart as astronomers, we would all have starved to death a long time ago.

A few happy accidents have advanced science because some patterns had been cleverly recognized and documented, even if scientists denied their significance. Absent those patterns, the unique observations might have simply been relegated to the same category where we put bizarre coincidences, like someone winning the lottery three times in a row.

Dr. Hynek often told me he'd waited years for the one case he could use to convince his colleagues, "running up the stairs of the Academy of Sciences," as he put it, waving that paper in the air with the definitive proof. That single case never happened. It wasn't Roswell, it wasn't the Kenneth Arnold sighting, and it wasn't Socorro. And if a nearly-perfect observation happened with obvious physical traces, it would only be significant in the scientific debate against the background of a well-studied, carefully documented pattern that everybody could weigh against their own statistical or evidential criteria. For that reason alone, I have been on the lookout for cases bearing similar characteristics to those of Officer Lonnie Zamora's report in Socorro in 1964 and those of Mr. Jose Padilla's sighting in San Antonio in 1945.

An observation that took place in southern France in 1965, fifteen months after the Socorro landing, is a case in point.

∧ ∧ ∧

The day was July 1, 1965. A 46-year-old farmer, Mr. Maurice Masse, got up early to work in his field in the "Olivol" area of the Valensole plateau, a beautiful region at the foot of the Alps, where grapes and lavender plants are cultivated, among other crops. Planning to advance his work before the sun was too hot, he was there at 5:45 am when he heard a whistling sound that puzzled him. His first thought was that an Army helicopter on maneuvers was landing in his field, and he angrily reacted with the intention of telling the pilot to go pick another place, in no uncertain terms. But when he walked around a hillock that shielded him he was faced with a strange machine (Fig. 29).

In his recorded testimony before Commandant Oliva, chief of the Gendarmerie at Valensole, Maurice Masse stated:

"At a distance of about 30 meters, I saw a strange machine the shape of which vaguely recalled a rugby ball. Its size was approximately that of a (Renault) *Dauphine* car, and it was of a dull color. It was standing on four sort of metallic legs and a central support. It looked like a monstrous great spider. On the ground, there was a human being of the height and build of a child of about eight. He was wearing a one-piece suit, but no helmet, and his hands were bare. Inside the machine I could see another being. Suddenly the one who was down on the ground turned around and saw me, and he immediately jumped into the machine. A sliding door closed behind him, and the craft took off at a staggering speed, giving off no smoke or dust, and in a fraction of a second it was all over, and the thing was out of sight."

The object itself may have flown off, but the traces it left in the soil of Provence were quite obvious. Those traces were going to be the topic of technical discussions in France for many years. The Advisory committee on the CNES-GEIPAN project (**40**) on which I currently serve in Paris has not forgotten the case, which remains in its files as one of the prominent "Unidentified" episodes that have resisted all attempts at rationalization.

The first reaction in a single-witness close encounter (*testis unus, testis nullus*, as Dr. Hynek had said) is to blame the event on hallucinations or hoaxing. But Maurice Masse was a prominent citizen in town (he owned and managed valuable businesses and his wife served as a town councilor; she was later elected as the mayor of Valensole) and he had distinguished himself as a fighter in the very tough Résistance groups, in battle against the Germans in the rugged terrain of the Alps during World War Two.

He was not someone you could safely accuse of a juvenile escapade in his own town.

In France, just as in the United States, the military first tried to explain the craft as a common piece of machinery. In Socorro, officials had tried to blame the sighting on an "Air Force Pogo," and later on a non-existing prototype of a NASA lunar lander. In San Antonio, Jose's father had been told the Army wanted to retrieve its precious "prototype" weather balloon. And in Valensole, the French military insinuated that they had maneuvers in the area, and Monsieur Masse had simply seen a light helicopter, probably an *Alouette*. Maurice Masse laughed at the cute "explanation," and so did the Gendarmes: an *Alouette* would have woken up the whole village, and nobody from the Army had even bothered to come over to the field and examine the marks they had recorded.

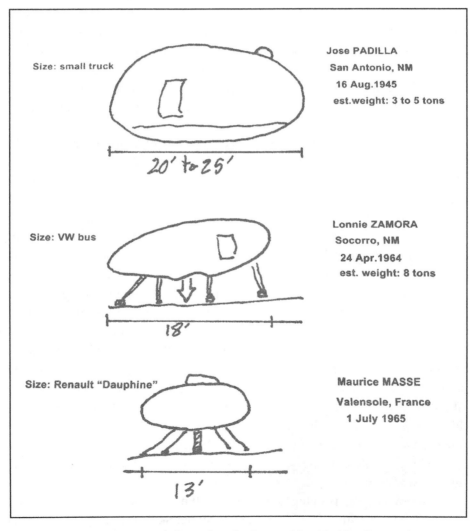

Fig. 29: Shapes and dimensions in three unidentified landing craft.

Chapter Sixteen - Global Patterns and a Surprise Third Witness

As we will see in the development of the investigations, these doubts and insinuations upset Maurice Masse and led him to withhold key elements of his experience from the police and from journalists. Other local witnesses then took the hint from their friend Maurice and understood it as a clear signal to keep their own observations to themselves, something the French do very well.

That situation is becoming a common one in the study of the phenomenon, particularly in the United States, where the most reliable witnesses are tired of being mocked on television by pontificating scientists with short-sighted agendas who have not taken the trouble to inform themselves about the facts, because they start from the assumption that they already know the answers in a general way, and only need to fill out a few details.

The Gendarmerie report signed by Mr. Masse, therefore, should only be taken as a first approximation, yet it is useful in stating some of the basic facts:

"We went to the scene of the landing," the Gendarmes wrote. "There is no doubt whatever that something has happened there. At the place where the central support was, you can now see in the ground a hole 20 centimeters in diameter and about 50 centimeters deep. And radiating out from it are marks in the shape of an "X" which would seem to confirm the description of the machine's four legs.

"But the strangest thing of all is that, all around the hole, the earth is as though petrified, hard as cement, whereas elsewhere it disintegrates and crumbles in your hand. But immediately after the machine had gone, the ground there had almost the consistency of liquid mud. And this is all the more strange, considering that it has not rained here for a long time."

∧ ∧ ∧

I arrived in Valensole on May 27, 1979, in the company of an active French diplomat who had the rank of Ambassador and happened to own a vacation home in that area of the Alps. We arranged to meet with Maurice Masse the next day through relatives of his in the village, and we began recalling his observation over drinks at a local bar, *Chez Dédé*, where much of the social life took place.

Mr. Masse told us that in his opinion, the beings were from "somewhere else" *but they were human like us*, although they were dwarves about three feet tall, and had oversized heads. Like Mr. Padilla, perhaps he found it necessary to think of them as "men" because they seemed to have expressions and behaviors that he could recognize, and with which he felt comfortable. Understandably, he did not focus on the differences at first, such as the enlarged head with respect to the body.

When I brought up the subject of the paralysis he had experienced at the end of his observation, a detail he had not initially reported to anyone, he didn't want to discuss it. **(41)** Nor did he want to discuss the spiritual questions the experience might have inspired. On an earlier occasion, he had shown my friend some vines growing along an old wall near his field, and said, "Look at all this life, it grows, it clings to everything, and then it will die and there won't be anything else. *For us, it's the same thing.*"

"So, had anything unusual happened since then?" I asked, knowing that serious witnesses often withheld such testimony. People know that repeated observations will decrease, not increase, the seriousness of the case in the eyes of skeptics, and it might even push the interpretation into those psychological categories of which American government scientists were so enamored, from "stress-related imagery" to "dissociative behavior." Psychology is a serious science, but the subject in this case had fought as a *Partisan* against German troops in serious mountain combat and he was a good observer of his surroundings. He was highly respected by the local Gendarmes who interrogated him.

It turned out that Masse had seen the object again: It happened at night, over another field that he owned. He trusted us enough, at that point, to feel free to discuss it, and he grew enthusiastic in his description. There were lights of many colors, whirling around; pretty colors; he compared it to watching the tail of some wonderful meteor. The colors changed so quickly you couldn't really see them: "When you want to look at the blue, it's already red..."

We spent the afternoon on the Oraison Plateau, where we inspected that site, and the next day we met again with Mr. Masse at his Valensole field, which he was irrigating. He told us that to see an object like the one he'd seen was "the best thing you could wish to anyone," adding that *a witness shouldn't say anything about it*, not even to his family: "One

Chapter Sixteen - Global Patterns and a Surprise Third Witness

always says too much:" One more point he had in common with Zamora and Padilla...

He also told us that he had started to think a lot about death.

∧ ∧ ∧

The records of the French Space Agency show that four other branches of the government took the initial Gendarmerie report very seriously and conducted investigations of their own. This included the internal and external Intelligence agencies, thinking of potential infiltrations of agents or other hostile events, and also the French Customs.

Why Customs?! I wondered.

When I asked the question, I learned they probably had the best reason of all to get involved: the region was occasionally used for financial contraband between France and Switzerland, particularly for gold. The *Plateau d'Oraison* was a very suitable reference and a safe landing site for helicopters.

All those investigations reached negative results.

Throughout our conversations with Maurice Masse—whether they were held in town, over a glass of pastis *Chez Dédé*, or at one of his fields—my friend and I had the impression of a very stable, emotionally and intellectually alert and clever man. But there was something more, the impression he'd been awakened to a range of perceptions or abilities he'd ignored before his experience, and he was still puzzled by it. After all, this was someone who had once told my companion that "we will die and there won't be anything else."

Was he reconsidering that certainty?

We spoke to his friend "Kilou" who said, "It's as if he now saw everything from the outside."

Masse himself put it another way, when I asked him about it. He said he was most afraid of his private experiences indirectly causing harm to someone else, as if he had become aware of external powers he'd never understood or felt before. He told us about a vision of his father's death, and other premonitory dreams that had taken him by surprise.

I brought up the question of any traces left by the short beings. That is one area he'd never described to government investigators, and he even joked about the letter of thanks the CNES Space Agency scientists had sent to the Gendarmes, expressing thanks for their collaboration: "I haven't told them half of what I'd seen," he laughed.

He'd also refused to be interviewed on television by Jean-Claude Bourret, a well-known media personality, and he never got involved in skirmishes with the press. He never argued with the skeptics.

There is a rampant assumption, throughout the history of ufology, that witnesses are always naïve and transparent, and truthful with authorities. That may be true in some cases, but when one touches on complex observations in a sophisticated environment like Provence, it takes more than a confusing questionnaire and a phone call from Paris to bring out the intimate truth about the witnesses' impressions, their doubts, and their awareness of changes within themselves. That was certainly true in the Padilla-Baca case in the intensely-secret environment of the atomic centers in 1945, and it also applied to an old fighter from the Résistance on the slopes of the Alps, where much unfinished social tension was still felt.

If truth be told, it also applied to Officer Lonnie Zamora, whose professional life was destroyed by the media frenzy, the resentment of his colleagues whose lives were perturbed, and the Air Force's obsession to promote ANY explanation for what he saw, no matter how fanciful or technically stupid.

All of which has led me to believe, over the years, that governments and their discreet contractors, even military Intelligence and the masters of aerospace, know a lot less than they think they do, or pretend they do, about the real phenomenon, its origin, and its designs.

∧ ∧ ∧

At that point of our discussion with Maurice Masse, I took a piece of paper and drew my best guess about the impressions left in the dirt by the diminutive pilots, as I thought they might be from some special cases in my own files. He looked at me in a serious way, made a few corrections, and the incident deepened the trust that had grown among the three of us over those two days: He volunteered that he no longer irrigated his fields at night, because he didn't care to see the objects again. Or their occupants. And he corrected a few points when we showed him his first testimony before the Gendarmes.

As we saw in the case of Lonnie Zamora, an important adjustment had to do with the distance to the object itself.

Chapter Sixteen - Global Patterns and a Surprise Third Witness

The journalists had reported that Masse was 30 meters away, or 50 meters away. The Gendarmes, at the spot, had estimated 60 to 70 meters. The truth, measured on the ground, was that he was 80 meters away when he *first* saw the object and the beings, but he'd stealthily moved forward through a vineyard (and not, as the media and the ufologists said, through his lavender field) to a point less than *six meters* away. That would make a very big difference, and here again we can see why technical discussions based on the press or even official records, in the absence of in-person, on-site analysis, can be completely misleading, whether they are based on a few pieces of paper or on a massive database on some computer network.

Applying artificial intelligence to the mess, as several groups are now proposing in the US and in England, would only be a huge waste of money, and more importantly, lead to false conclusions, in the absence of several years of very tough scrubbing by experienced teams.

Similarly, the press articles state that he saw the object and the occupants for 90 seconds. That's obviously not true, but Maurice Masse did say that in his early testimony, probably to minimize the importance of his experience and the pressures placed on him, since he had little trust in his interviewers.

He had seen much more, and possibly too much; he hadn't had time to process it all, so he decided to keep the actual truth to himself. Again, the reaction of a crafty Partisan.

He had lied, too, about the small beings, saying he saw one of them inside the object. Why? He may have thought that might make the story more believable. If both occupants were out in the field, they must have had a specific occupation there, and a subtle relationship with him that he wasn't ready to discuss.

Maurice Masse had also lied about the two beings when he told the Gendarmes they had hair like us, as he compared their size to children, who are never bald. In private, he told me the beings had bald heads. We have seen the same hesitation in Mr. Padilla's responses to Paola's questions, when he keeps talking about "the men," while unconsciously resisting the details that might indicate that the creatures were not quite human. Or perhaps, not human at all?

The main "detail" Maurice Masse withheld from everyone for several days had to do with the paralysis that froze him to the spot when one of the occupants pointed a small device at him before flying away. Once the object was gone, he remained glued to the spot for twenty minutes, standing up but unable to walk away, staring at the empty landscape, thinking he was going to die there, unable to call anyone.

After a while, those secrets he kept were too much: His wife confessed to my companion that after a few days, Maurice had burst into tears and confided to her everything about his observation. But there was one point on which he'd never publicly told the truth: Not only about the fact that—as in Socorro—the two beings had left footprints, but the fact that *he went back to the site alone and erased those footprints himself.*

When we met with him that day, he told us he'd noticed the same footprints at his other field, a sign that they'd come back. We had the impression that this had become like a private signal from "them" to him, and that he wanted to keep it that way.

∧ ∧ ∧

Maurice Masse, for reasons of his own, was reluctant to discuss the other cases in the area. In that sense, he was protecting his local friends from the kind of absurd publicity to which he had been subjected, and the sneering attacks from academics: Given the media circus triggered by the initial reports, and the steady barrage of furious "explanations" from scientists who hadn't bothered to travel to Valensole or to consult the police records, everybody around the area with relevant information had simply clammed up.

The French countryside can be very closed-mouth about local happenings, a social trait that reflects centuries of distrust with government and with authority of every kind, and gives the French village life its special flavor. (There are many savory jokes in France about the plight of outsiders trying to ferret out any truthful statement from local farmers).

This is especially true in Provence, once a wholly separate nation from the arrogant and vengeful France centered in cold, nasty Paris: the tensions of the previous decade that gave rise to bloody local conflicts and almost to a civil war at the *Libération* from Nazi rule were not totally forgotten.

Again, what we saw there was a juxtaposition, if not a permanent undeclared conflict, between two very different cultures. Maurice Masse would not tell us anything more.

It was Kilou, who had become our friend over those two days and knew my companion from earlier visits, who told us about some of the other cases, and his own experiences, which had never been reported to anybody. For example, there had been a case four years later, in October 1969. There, the leaves of two trees close to the landing site had instantly turned yellow, and traces were left in a wheat field where nothing would grow for the next few years. We visited the place. The farmers told us the traces had been in the shape of an oval, nine meters long, like the scene Paola had found on her first visit to the San Antonito site with Mr. Padilla.

In another case where we also met the witnesses, a farmer and his team had been busy in a field, operating two tractors, one with a Diesel engine and one with a conventional engine; an unidentified object flew over at close range and the conventional tractor lost power.

The one with the Diesel engine kept running.

^ ^ ^

A final word concerns the "paralysis" that affected Maurice Masse. Calling it by that term is a misnomer: If Masse had been subjected to a paralyzing weapon, he would have fallen down and his heart might have been affected.

What actually happened was incapacitation, not paralysis. It was triggered when one of the beings took a small device from his belt and pointed it at him as he emerged from the vineyard and faced them, still assuming the craft was some sort of French Army prototype. As I've mentioned before, it was a form of nervous system interference where the victim did not fall, although he was unable to move.

When researcher Aimé Michel published these details in a specialized British magazine (**42**) he speculated that the effect might be caused by some form of radiation affecting the reticular formation in the brain. But neither Aimé nor I have a background in medicine.

The article attracted the attention of an English physician, Dr. Bernard E. Finch (**43**) who found the idea unlikely: The reticular centers, he said, "control the central services, i.e. breathing, heartbeat and muscle tone, etc. Injuries to parts of the reticular system are liable to cause tremors, involuntary movements and various rigidities."

What Maurice Masse experienced, however, was more adequately described as temporary paralysis of *voluntary* muscles, Dr. Finch wrote, and that would be better explained by "a force field affecting the motor cells situated on the outer surface of the cerebral cortex...along a line running from the ears on either side to the top of the cranium. The sensory area is behind the motor area. The force field therefore produces its effect on all the surface cells of the cerebral cortex, and its effects wear off as it penetrates into the brain tissues, disappearing a few millimeters inwards. This dampening effect is produced due to the thickness of the brain tissue, and as the more vital centres are deep in the brain, they are not touched by the field at all."

Why would this explain what happened to Maurice Masse?

Because, Dr. Finch went on,

"Surface cells receive the brunt of the field, resulting in spastic paralysis and sensory changes involving touch and feeling, and sight. One can only speculate on the mechanics of the field proper on the nerve cells. Being of a magnetic nature, the insulation of the nerve fibers is broken down, nerve (electric) currents will not flow, the cell does not function, and fills up with granules that normally appear in periods of intense activity."

In the three days following his observation, Maurice Masse did not notice any particular effect on his sleep patterns, but on the fourth day he suddenly collapsed, seized with a need to sleep that he could not resist. Dr. Finch proposed to explain this by the long healing time needed by the cells of the cerebral cortex to recover from the intense force field, hence the sleep and tiredness.

He concluded:

"This would also explain how reflex actions, which are motivated by deeply seated cells, are still active in spite of the force field."

The weapon, he went on, probably aimed a concentrated beam at the head.

Note that the above was written by Dr. Finch in January 1966. The study of the brain has made giant strides since then, with better interpretations of what the witness experienced, but it is useful to have this initial explanation in the context of the time, when brain control weapons were still bulky and on the drawing board.

One should keep in mind that the 1965 device held by the "occupant" was the size of a cigarette lighter... and had no visible power supply.

∧ ∧ ∧

We now have sufficient data for a fair comparison of the characteristics of the entities (or "occupants") and the objects described in these encounters.

Chapter Sixteen - Global Patterns and a Surprise Third Witness

1. In all three cases the beings in question breathed our air without any devices.
2. In all three cases they measured about three feet, or 1.2 meters, the height of an average eight-year-old child (range: 47 to 54 inches).
3. In all three cases they wore coveralls but no helmets.
4. In all three cases the witnesses felt unafraid of "them," or even reported a sense that "they" had good intentions towards us, although they couldn't ascribe this impression to anything in particular.
5. In all three cases there was a psychic impact in the form of visions, dreams or direct "telepathic" impression (recall Jose's dreams of "clouds falling").
6. In all three cases the witnesses knew far more than they ever reported publicly or to authorities, and they shied away from publicity to the extent they could.
7. All three cases seemed to suggest an outer-space connection or a spacetime connection, compatible with the state of our own human culture following World War Two. They displayed behavior, instrumentation and engineering mastery recognizable and (almost, but not quite) compatible with human concepts of the time.
8. Normal vegetation did not grow back at the two sites of San Antonio and Valensole.
9. An interesting, large fungus was stimulated in San Antonio, as in other unexplained cases in the literature.
10. In all three cases the object was described as having an ovoid shape, rather than a disk.

There was no classic "flying saucer" in any of these testimonies.
In fact, there is no flying saucer in this entire book.

Now for some of the differences:

1. In the Padilla-Baca case, the three beings were able to move by instant linear motion.
2. In the same case, they bore some resemblance to oversized insects, like some crickets.

3. In the same case (a crash) they seemed disorganized, unable to take control of the situation.
4. In both the Socorro case and the Valensole case, the encounter seemed to be a set-up, a staged event, where the beings pretended to be busy but were actually manipulating the perceptions *of a human witness they were expecting*. They remained in control throughout, and left the scene easily.
5. The form of apparent propulsion seems to be an elective: It varied among all three cases. In the 1945 case, no propulsion mechanism was evident, although the craft appeared to be steerable as it fell. At Socorro there was a terrific noise and a blue light but no exhaust (all noise and other effects stopped once the object hovered at a low altitude). At Valensole the craft lifted silently, then it just blinked out after a short horizontal distance, with no sonic boom or other physical effect.
6. Valensole was the only case (of the three) where a weapon (non-lethal) was used.
7. Valensole was the only case with overt physiological effects (paralysis, followed by a pattern of sleep disruption).

To most American ufologists, who insist on taking such observations strictly at the first degree ("If it looks like a spacecraft, it must be a spacecraft"), these features must be a proof that we are visited by travelers from somewhere. Then, they argue, if funds were allocated to better capture the information and study the traces left by the craft, we could learn important facts about space travel and perhaps medicine as well.

That is a reasonable reaction, but one shouldn't derive too much from such simplistic hypotheses. In America, however, our research culture is such that one must present a practical hypothesis to be tested before anyone will award you a contract or a grant, so this has resulted in almost no fundamental academic research on the subject since 1945.

In the rest of the world, there has been considerable theoretical speculation and extensive documentation of events, with a superior appreciation of the importance of historical reports, but the organizational reaction has been similar, with no incentive to fund active research in the field, other than government-controlled and government-censored teams of supervised volunteers, like the French group under the CNES space

Chapter Sixteen - Global Patterns and a Surprise Third Witness

agency, or very specialized, necessarily narrow-focused classified studies, starting with collected hardware.

To us, it seems that the phenomenon (and in particular the three detailed cases we have reviewed here) functions with reference to a more complex strategy. It displays just enough to intrigue the witnesses and the society around them, and it uses imagery recognizable to us, but it is sufficiently flexible, in time and in space, to defeat simple analysis.

If that is true, then it's unlikely that we will learn anything really useful or operational about the phenomenon as long as our studies, overt or secret, only concern themselves with the overt technology on the scene, as staged for our benefit, and framed for our physics.

There might be another approach one could take, if the people who have some of the best physical evidence would come out of the basements where they have been hiding since 1945—the people who came to retrieve their "weather balloon" on the land of Mr. Padilla, and confiscated the "ordinary silica" from the material Mr. Stanford had submitted to them.

We believe the phenomenon is not trying to teach us physics or advanced propulsion, although most researchers will keep trying to extract such data, or turn it into new weapons. Instead, we believe it is trying to teach us to transcend our own humanity. Or perish in a toxic mental cloud of our own making, when our civilization reaches its point of singularity.

^ ^ ^

Time passed. The witnesses grew up into adulthood, got married, had kids of their own.

For Reme Baca's wife Virginia ("Ginnie"), Earthlings were the center of the universe and the word UFO had no place in their home, as she recalls in *Born on the Edge of Ground Zero*: "It didn't matter how many times Reme brought the subject up, I would not acknowledge it. There was no such thing. If there was, God would have let us know through his teachings, after all."

By 1994 they were living in Tacoma, Washington. She had visited San Antonio with Reme and was pleased to note that one of the largest Catholic congregations in the county was located there. But something was going to alter Virginia's view of the world on a Sunday evening that July, although it wouldn't change her faith. By then, she says, "Reme had made up his mind that *it was not that important for him to share his experience* of having discovered the wreckage on an alien craft with society, *it was no longer an issue with him.*"

On that particular evening, in June 1994 about 10:45 pm, after picking things up in the kitchen, she joined Reme outside; they enjoyed looking at planes coming in for a landing at the nearby base, under clear sky, not a cloud in sight.

The craft came in from the direction of the Narrows Bridge, and made a deliberate turn towards their house.

It was now above them.

It covered the sky, and they were underneath it. A long silence followed.

"There was this eerie feeling that brought back memories from when I was a kid," Reme said. "When Jose and I discovered that crashed object near San Antonio, *there was a silence in the air. It appeared that the world stood still.* Now it seemed to be happening again."

"You couldn't find a person to save your soul on the street that night," Virginia added. "Dogs quit barking. Things didn't seem right. There was this deep bright orange lighting that appeared to be imbued in the underside of this giant craft. The lighting seemed to glow rather than shine (...) The colors seemed to be projected from inside the craft or emanated from it, instead of reflecting as from a painted surface or a mirror.

"There was no noise at all, no sound. The craft just seemed to sit there as if it had been painted into the sky. It was totally amazing. It was there for way over fifteen minutes; I know because I checked the time when we went back into the house, and it was well after eleven."

Suddenly, the craft began to move: slowly at first, then at increasing speed. They could see what appeared to be a dome on top of it. It picked up speed until they couldn't see it any more.

Virginia went to lay on the couch in the front room after that experience; when Reme returned he found her crying, repeating "God never told us what other worlds He created."

When Reme spoke to military personnel at McChord Air Force Base, they told him they weren't taking reports of "Bogies" any more, but they referred him to the J. Allen Hynek Center, where Paul Hynek took down the data, and told him that his father had died.

∧ ∧ ∧

Chapter Sixteen – Global Patterns and a Surprise Third Witness

On Friday, the 16th of October 2020, Paola and I were back in Socorro one more time, to meet again with Mr. Padilla, to better understand the site and work on details of his observations. The weather was fine, the sky a perfect blue at this altitude, with a bit of steady wind, shirt sleeve weather. Jose was recovering from an operation on his first bullet wound, the one from Korea. The second wound would require another surgery, a few weeks away. He now walked with a cane but after a period of depression his old energy was returning. He clambered up and down the rough terrain ahead of me.

The dignity of this man was an inspiration.

When Paola asked about his plans, he said that at 84 he was spending more and more time at the cemetery, where he'd just paid for a new grave marker for Faustino and made preparations for his own place, next to his parents. He suggested we visit the site because from up there we could see the whole village, so I drove our rented car up the hill on a dirt road to the plateau where the newer graves were clearly outlined in freshly painted white among the wild flowers and the stubby, low bushes growing free. Indeed, from that spot, the limitless views over the valley were magnificent, the silence and the occasional flight of a bird of prey high above us an appropriate setting: The stark memory of the terrible, festering wound of the bomb hovered there, the product of a time of worldwide passion and collective trauma in the name of science that only future history would be able to solve; and absolve?

What had brought us there was the evidence of the crash of an extraordinary vessel with extremely weird passengers, and of the week-long, well-observed recovery of the evidence by the United States Army. Some of which we had in our custody, thanks to a couple of irreverent urchins.

Evidence of what?

What a funny word, "evidence," I suddenly realized. We spoke of it so casually...

We needed to keep filling the blanks in the back story, and Jose was eager to help with the research, so we drove back down the slope to spend time at the site of Faustino's old house. We also went to the sheepherder's shack where the strange creatures had made an appearance, now just a picturesque ruin of adobe walls and twisted trees. We saw Reme's old home, nearby. Then it was time to climb back up to the crash site, survey maps in hand, to complete the review of what we'd missed on prior trips.

Where the Marconi tower used to stand there's little to see now; a casual visitor would find no trace of it. But it comes back to life when Jose describes how he actually climbed it and checked the bent metal girders where the craft had hit. He led us around the perimeter and located the areas where the four legs had been anchored.

Given the distance covered, it must have been a massive tower.

From the high spot where we stood, the three peaks of Trinity were clearly detached on the horizon. That brought back some memories of the nuclear blast.

"My mother told me about the big light, and we both got blinded in one eye when we stared at it," Jose said as he looked over White Sands. "She never regained sight in that eye. I was luckier than her, I recovered my vision, but when the terrific sound wave came, my right eardrum ruptured. Blood ran down my neck. When Faustino saw that, he thought I'd been shot."

Did the craft come from Trinity, as Jose believes because of the location of the torn metal as the device bounced off the tower and went on to crash by the dam? And what about the silvery spider web spilled from the break? It was all over the bushes, he said. The kids were fascinated, they gathered it in a big sack.

As we stood there I glanced to the west and saw the tiny outline of large animals on the far ridge, so I called Jose's attention to it. It was the first time in two years that we'd seen cattle there. So we drove on to the crash site and found a group of a dozen black cows next to the new pond, and one white cow on whose shoulder I saw the vague outline of a brand.

^ ^ ^

On the way back to our Socorro motel, I naturally was reminded of Dr. Hynek's investigation, so I asked Jose for directions to Park Street and I drove up the dirt road Lonnie Zamora had taken to the landing site, back in 1964. I had expected a lot of changes in the landscape, but it seemed very close to the old descriptions from writers like Ray Stanford, whose reliable hand-drawn map I used, and from Allen himself. If anything, the roads had become worse, rocky as Hell, deeply eroded and partly washed out by many rainy seasons. Jose and I scrambled up the small hill beyond the new Baptist church to inspect the terrain.

So I'd finally closed that loop, and knew I could stand more firmly on what I'd written.

Chapter Sixteen - Global Patterns and a Surprise Third Witness

I had dinner with Paola. Inside service had been banned, so we sat in the courtyard of the restaurant and we spoke of our lives, the people we knew, and the complicated games officials were playing. I showed her the red eye of Mars staring at us, high in the pure sky of New Mexico. The planet was at its closest approach to the Earth in many years.

In the classical New Mexico décor of the Socorro Springs Brewery, she shared with me the conflicts she felt, and why this particular story was so important to her. We didn't record the conversation, but it struck me deeply enough, and it's still recent enough, for me to reconstruct it.

It went like this:

Paola: This is one of the most important events I've ever investigated. I've shared this case with many researchers, and none of them took it seriously because they didn't understand the technical aspects, and the gravity of this particular case.
Jacques: Because the witnesses were so young? They wouldn't trust them?
Paola: Children don't lie, certainly not in this way, about such an event. Every psychologist will tell you that.
Jacques: My late wife was a child psychologist, remember? Janine told me that often. She'd agree with you.
Paola: These children had no reference with which to compare what they saw. It ruined their lives! It has left them without a resolution, other than telling the story to us as accurately as possible for us to save it, as we're trying to do. I treasure the interesting dialogue we've had over these three years. It's personal. Psychologically, it's important for me, to make some sense of this strange reality.
Jacques: You've already dealt with that reality in your previous books, though.
Paola: Not quite this way; although gathering witness testimony with military and intelligence personnel, I did maintain many friendships. It's never just about data collecting. It's also about sharing and empathizing with the witnesses about the nature of the phenomena. What disappoints me is the lack of collaboration among researchers, even the most dedicated ones.
Jacques: The field has fractured into many little discordant voices, from hoaxes to grand Theories with no relevance to actual cases…That's what the public sees. The fake stories on cable channels are just as

entertaining as the real ones, sometimes more. People react with their emotions, not their brains. And scientists are the worst!

Paola: It's all become primarily an entertainment media. If we all could share research and work together like you and I are doing now, Jacques, think of what could be discovered! We could bring in the ideas we value most, input from specialists in an open forum: the metallurgists, the medical professionals. I've rarely had access to them, and they rarely work together.

Jacques: They've never taken the time to investigate cases like this one. It's complex.

Paola: Not really, it's not complex. It's simple, and it's important. It belongs to humanity, not to you or me or the government or someone at Google. It's real. It happened. But you need a combination of different perspectives and the ability to bounce off ideas with others, as you and I have done. That's where the complexity lies, the integration.

Jacques: You and I have dealt with this all our lives: doesn't that help?

Paola: My whole life work has been to capture the *exact* words of witnesses, but that hasn't led me to any conclusion about the mystery. Instead, my focus has been on historical accuracy, even if individual perceptions were distorted. But this case is unique: It moved me into another realm, to try to understand the meaning of reality, the reason for the manifestation, and the impact it had on the lives of Jose Padilla and Reme Baca. They are our witnesses, our experts, our explorers.

Jacques: But listening to them…That in itself gives you incredible insight.

Paola: I'm not so sure. I've never really examined what it could all mean. Maybe I should trust in my own ability to make some sense of the whole mystery, based on what they tell us. But it's true this case has created a new perspective for us both because it involves children, it also involves Native American history, it involves the government, it involves scientists engaged in deadly top secret projects, and it involves a time where humanity shifted into a dangerous era.

Jacques: A big transition: The Atomic age. Everybody takes it for granted now. Ancient history.

Paola: But they're wrong! The Atomic era is still relevant. We're still playing an international game of nuclear chess where every winning move leads to *checkmate.*

Jacques: Perhaps, but it's more complex than chess, because of all the levels of secrecy. It's obvious the Intelligence systems are confused by all

Chapter Sixteen - Global Patterns and a Surprise Third Witness

the fault lines in their own abstruse classification systems. The UFO problem has made that clear. Reme Baca knew more about this crash than the director of Central Intelligence, swimming in ambiguous signals of his own design.

Paola: I've always been comfortable living in ambiguity. My goal has never been to answer all the questions from the general public. The more I find out about this reality, the more questions I have! We have that in common.

Jacques: That, and the underlying reality of this beautiful place: pinkish blue New Mexico sunsets and the freedom of the high desert. A great setting for an enigma like a UFO crash!

Paola: Yes, New Mexico has its own mystique. In my travels, I've gone to White Sands Missile Range where Colonel Philip Corso had worked in 1957. I even went to the infamous Holloman Air Force Base. But every time I come to Trinity, Stallion Site, the site of the atomic bomb blast, as we did, I have the same intense emotion viewing the Oppenheimer statement from the Bhagavad Gita on that roadside marker: "I am the Killer of Worlds." *Notice the plural! Not just this particular world...All the worlds...* I feel profound respect for the power of scientific developments, but I dread the possible implications.

Jacques: Speaking of implications, we have to think through what this book will mean. We're opening up a new inquiry in San Antonito but it could affect our consciousness and open up new public awareness of the real problem...or...

Paola: ... I know what you're going to say: It could just as easily become the next UFO circus. That's what it's all about, isn't it?

Jacques: Let's make sure it doesn't turn into a circus, as Roswell did, with all the hoaxes, all the exploitation by every group, including the blatant lies by the military. It has been so hard for the real witnesses to have their story heard! This case, with the testimony of Jose and Reme... this case is unique. You're the one who's responsible for revealing it, and documenting it.

Paola: All I know is that it will stay with me for the rest of my life. It's probably the most important investigation I've ever done, perhaps the best contribution we can make. The irony is that I'm passionate about giving a voice to the witnesses, telling the story of Jose and

Reme. What happened on August 18, 1945 should be remembered. But it could start the same complex reactions as Roswell did.

I recalled that 1945 was also the year of Paola's birth. She commented on the irony of beginning her research career with Dr. J. Allen Hynek and continuing it with me. "There's something metaphysical about all this," she added. "Unless it's just a big simulation game in somebody's super-computer in the sky," I said, "as some of my AI colleagues suspect."

∧ ∧ ∧

Going back to that hillside the next day, I saw that Paola had bought flowers for the family graves (why didn't I think of that?) and we spent time listening to Jose as he reminisced about his childhood, his days as a law enforcement officer in California, the time when he was shot, all the changes in his family, "back here," while he was gone.

The Corona virus, whose decreased force during the summer had been deceptive, was winning a renewed battle in America. Bitter electoral politics was clearly to blame, splitting the country in nasty ways. But that wasn't the whole story: European statistics were actually far worse, and climbing. It was a planetary fight now. I thought of my children and of this new period of danger coming upon us. I felt sad, and scared. When I reviewed the pictures I'd just taken at the cemetery I realized they told a story so intense, of life and death, of centuries gone by, of history hidden and rediscovered, that the experience shook me.

Walking around the area, I discovered a burial marker from the 17th century behind a clump of mesquite. There was another isolated marker, very interesting, obviously carved by a local craftsman, which combined an Aztec motif with a Christian cross. I found myself, once again, in awe of the majesty of that landscape. I was beginning to grasp how one could become emotional in an open desert.

Time flowed differently, too. This was not like a prettified picture in a magazine or a few frames from a documentary on the Web: It was real. There were traces of vibrant lives, long ago, that oddly remained. You couldn't understand anything about the present, standing there within a few miles of the site of an atom bomb blast, without reconstructing, documenting and analyzing the tiny details of the deeper past; and it was an open book, if only you troubled to read it.

Chapter Sixteen - Global Patterns and a Surprise Third Witness

Even Ground Zero was rooted in rich layers of memories the scientists hadn't had time, or taken the time, to recognize.

The desert wasn't empty. People may have passed on, the Tribes may have migrated across the mountains to Arizona, but you could still interrogate the landscape and get answers, if you only attuned your mind to the scale of things in the well of time.

That started another line of practical thought: might some local people remember the bright stuff Jose and Reme had pulled from the bushes at the crash site? And what about the funny metal you couldn't bend because it always straightened itself out, mysteriously?

Mr. Padilla was skeptical about that search: The spider web had been widely distributed to his neighbors in San Antonito, he said, but in later years, as the modern world asserted itself, people wouldn't have used it for Christmas décor any more. Families moved, stuff got thrown out. Some houses had burned down, or had been torn down in the "modern" era.

There was someone, however, who might remember something, Jose said. He had a niece, younger than he was, a girl from another branch of the family. When things had been tough, a few years after the war, Faustino Padilla had given her a home and raised her like his daughter. She might remember all the bright stuff the older kids played with.

After all, nobody else in the area had super toys like those.

Paola didn't miss a beat: her investigative instincts took over. Back home in Colorado a few days later, she picked up the phone and found Sabrina.

^ ^ ^

When Sabrina answered, the conversation began with Paola introducing herself, and explaining why she was interested in those events everybody else had long forgotten.

Paola: I need your help. That's why I'm calling...
Sabrina: Oh, OK, I hope I can help you.
Paola: I have to ask your age, because I'd like to take your statement about what you saw. So, I have to ask your age.
Sabrina: OK, 67. I'm, um, I just said, these years are just going like this, you know, but uh (*Chuckles*) what can we do? You know, we get there, somehow.

Paola: Well you're 67, so when you were living in San Antonio, how old were you?

Sabrina: Oh, gosh, when I was there, I went to San Antonio Elementary School. That was near the church over there--they had like a Catholic church there, which we all made our Communion there. But I didn't go to high school there. I went to high school out here, in California.

Paola: OK, and what I need you for, Sabrina, is I need you to remember if Faustino took you to the field, near where all that stuff happened.

Sabrina: Yeah, I've been there before, but I was little when I went and I saw it, uh, it was a terrible sight, you know. It was very scorched and, yeah, it was a terrible sight.

Paola: So, how old, can you give me how old you were?

Sabrina: I think when I saw that, when I went over there, it had been some time already. They had already took the, you know, they had carried everything away and all. Everything was practically gone already. But what I remember seeing, when I walked up there by myself – cause you know I was kind of, I wasn't supposed to even…be over there. But you know kids. You know I was doing things.

Anyway, I walked up there, but I didn't go all the way. I could see it from where I was standing. It was a good walk, 'cause it's like a rough terrain in there. And I could see that the ground and the whole terrain there was scorched black. I remember that. And I remember seeing a tree that used to be there. And whatever burned up there burned up so intense that it left the tree…you could just see the skeleton of the tree, where it just….You know what I'm trying to say: when something burns up in such a quick fashion it might even leave like a skeleton behind, you know, like the skeleton of the tree was just still there. It just looked eerie, very eerie.

Paola: But when you saw the scorch, was it in a shape, or was it just all scorched?

Sabrina: Uh, it was all scorched there, it was a big area that was scorched. It was a huge area.

Paola: Ok, and was it in any particular shape, or was it just burned all over the place?

Sabrina: Uh, well at that time it was a big area to me, 'cause, you know I was little, of course. It just looked like it was bigger than a football field, I would say.

Chapter Sixteen - Global Patterns and a Surprise Third Witness

Paola: Ok, that's really interesting, 'cause the area today is a circle, 'cause things grew back.

Sabrina: Yes, uh, probably a lot of things have grown over it, but I think my Uncle (*Jose*) told me some time ago that a lot of things didn't really grow out there good. It was, like, things didn't hardly grow back, in that particular area where there used to be trees, and well, not, you know, certain trees, and plants and things, you know, because he had these little ugly plants. I used to hate them things, I remember them.

And he said it was strange how nothing would grow there, and then something all of a sudden started growing there and I said well, be careful because, after all, you know, we don't know what's still there, that you might get some kind of disease, or something, and you know he's an older guy and...

Paola: Yeah, I need to know about how old you were, though?

Sabrina: I think I was about eight-years-old, seven to eight-years-old, (*note: in 1960 or 1961*) when I walked up there by myself. I wasn't supposed to have been over there. It took me a long time to walk over there and back 'cause, you know, you gotta go through rocks and a lot of rough terrain, and I got scared when I saw that, because it almost looked like I was going into a black forest, the way it was scorched and everything. It just frightened me and I turned around and I came back home, but I remember that part.

∧ ∧ ∧

That informal introduction already told us a lot: Evidently, Sabrina had retained a vivid memory of her days in San Antonio and her view of the site. Her descriptions, so spontaneous and in such direct responses, were rich in suggestive images: the lack of growth, the blackness, the wounded earth, the "little ugly plants" she hated as a little girl.

And something about the site frightened her. That was the point that Paola pursued:

Paola: Did they tell you what happened? How did you know, that you weren't supposed go there? How did you know that?

Sabrina: Well, I used to walk up there all the time because, you know, I was the only child there. There was no one else, and I used to go up there to see the cows and stuff, and the horse. We had a horse there at one time and I used to walk up there because they used to have, like a little hill where they used to...some of my family members would come and they would do target practice, all the way from our house. And then they would have like a hunting season. So, what they would do, they would go hunting for deer...

Paola: But how did you know you weren't supposed to go there? Did Faustino say, "Do not go there"?

Sabrina: No, I had just wandered over there. I had wandered up there a couple of times, that area that was scorched like that, and I just came upon that ...it was scary to me where I didn't want to continue, it just looked...It just didn't feel right to me, you know. And then, you know, being a kid, you get these feelings, like: "Oh no!"... (*chuckles*)

Paola: So, you just had a *feeling* you weren't supposed to go there, right?

Sabrina: It was just... I walked over there on my own. I just wandered over there 'cause I used to walk to the cemetery all the time, by myself.

Paola: I mean, they didn't say, "There was a crash, we don't want you to be there"...They didn't say anything to you, or did they?

Sabrina: No, *they never told me that*, because at that time, they had already took most of that stuff out that was crashed, and everything was gone, but it wasn't too far from where they had a cemetery up there, and I used to go to the cemetery. I was used to walking around that area, where the cows were. And it's a little walk.

At that point Sabrina's testimony was beginning to fill the gaps we had noted, regarding the post-war era in San Antonio, and it all made sense: the dying vegetation, the scary appearance of the landscape after the crash, although it remained unclear for us whether the burned area was a result of that episode, or an unrelated lightning strike and fire.

Recall that in Chapter Two, reviewing the testimony, we had noted Reme saying that the mesquite burned when the crash happened.

Sabrina's remark, that vegetation has begun to return to somewhat active growth, was interesting. We did know, however, as any visitor to that part of New Mexico can observe, that plants remained very weak and the large trees used by the Army to camouflage the craft in 1945 had never grown back.

∧ ∧ ∧

Chapter Sixteen - Global Patterns and a Surprise Third Witness

The next exchange was about Paola's probe about unusual samples or pieces of material retrieved from the area after the recovery of the craft and the departure of the young soldiers.

Paola: Did you ever see any metal or pieces or anything yourself?

Sabrina: Well, I seen the metal pieces that... when my Grandfather (Faustino) had these strange little pieces or strips of metal at one time. I remember I was little and we were laughing. I remember we were in there, in the kitchen laughing, 'cause these were little strips of metal...*There were a couple of them*, and they weren't very wide and, you know, we would just fold them up and crush them up, and the strangest thing: it would just flatten right back out!

I don't care how you fold it up, and how you ball it up--you know how you ball up a piece of paper and you say, "Oh I made a mistake, I'm gonna throw this away?" You ball it up and throw it in the trash. Anyway, every time we would crush it up, ball it up, or whatever, it would straighten right back up like nothing happened! It was a strange....

I remember I played with it until I got tired and I think my Grandfather probably, I don't know whatever he did with that. You know, he might have thrown it away, thinking, "Well, what could he possibly do with that?"

Paola: So, your Grandfather Faustino had it. Where did he keep it?

Sabrina: I don't know. I just know that he had it at that particular time, and I got a chance to touch it and I know I saw it. I just thought it was like a joke, almost: you know, you pick this thing up and it doesn't stay the way you leave it: It straightens back out.

Paola: How old were you when you were playing with this stuff?

Sabrina: Oh, maybe about, let's see, maybe about seven or eight-years-old, that I can think of.

Paola: When were you born? When's your birthday, your whole birthday?

Sabrina: March 27, 1953.

Paola: So, your Grandfather had that for at least, oh my God, seven years he had that stuff.

Sabrina: Well, I don't know how long he had it. I just seen it and uh... you know at that time, I just said to myself, well, to me, "What could you use something like that for, to even save something like that?" You know, like, when you find something, you say, "Should I keep this or throw it away, or give it away?" But I don't know whatever came of that.

    I do remember the thing we used to call it, "Angel Hair", but my Uncle used to call it "Spider Webs."

Paola: You do remember that, you do remember the spider webs?

Sabrina: Yes, I touched it. I touched it. And to tell you how it looked, what color it was, 'cause my Uncle has bad eyes: it was like how old women, when they get their hair gray, they put this color in it and it's like a cellophane, like a purplish.

Paola: Yes, my grandmother used to do that.

Sabrina: Yes, well, this "hair" had purple in it. *It was purplish, like a blond purplish.* And it was shiny *and it would glow*. And I remember I played with that many a times. My Grandmother had it in a little bag, in a little plastic bag, and I don't even know why did she have it, but anyway I touched it many times. But I never played with it long. And I'm gonna tell you why. *Because it would hurt. It would cut your hands.*

That observation was sensational.

Sabrina, who doesn't know much physics, had just told Paola something that changed the course of our investigation.

Not only did she describe the same strange "memory metal" that would be reported at Roswell two years later, but she was telling us about something else: bundles of fiber that glowed brightly enough, seven years after the crash, to be visible in the dark. One of the possible conclusions (which we couldn't verify without getting some of the material) was that it was radioactive; and that had implications for the nature of the craft: If it was really an Alien vehicle, which came from far outside New Mexico, why would it have been affected by the atomic bomb? Or were there other hypotheses?

That didn't make sense. Checking this would be a high priority.

Paola: Do you know if there's any of that left anywhere?

Sabrina: Well, I doubt it because since then… we had the house back then, it burned down and some other people got it now. But I know that she used to keep it in a plastic bag; and I think I know now why she kept it in a plastic bag, because it was strange. I mean, *it would even*

Chapter Sixteen - Global Patterns and a Surprise Third Witness

glow in the dark. You know, I remember when we used to have it in this other bedroom.

It was always dark in that bedroom, and it would glow in there, I remember that, 'cause, you know, we barely had electricity back then. We didn't have indoor plumbing, and uh, that particular room was always...

Paola: So, you remember it being in a plastic bag and you remember, now you call it "Angel Hair." Your Grandpa used to call it "Spider Webs". What did you guys do with that stuff?

Sabrina: I don't know. I know my Grandfather probably had more of that stuff in a bag or something. You know, when you think about it, when you're a little kid and you see something like that, and you think that's something that you didn't want to handle too much 'cause I got, you know...when you touch it, it takes a little while when you touch it, it would cut your hands.

It was like a bunch of little razors touching you, I mean cutting you, but at the same time you wouldn't be bleeding from it. But it was very uncomfortable, it was almost like it didn't want you to touch it. I don't know how to...

Paola: No, you said it very well, uh, but Jose, your Uncle, says he used it for Christmas decorations.

Sabrina: Yeah, they used to give it to people and people would put it on Christmas trees or, hang it up around, or whatever. 'Cause I think they had so much of it, they had it in oh, uh, maybe a gunny sack or some kind of bag that they threw in the back over there, 'cause, mind you back then, you know we didn't know, these things, what it was used for, what can you use that for?

The idea of the kids picking up bundles of fibers from a crashed vehicle that looked like a weird spaceship, and giving them away to their friends and neighbors to decorate Christmas trees, boggles the mind just a little. We got used to it after a while: of course the kids would do that! They wouldn't carry that stuff back to Los Alamos; or mail it to the Pentagon.

But the conversation with Sabrina wasn't over.

^ ^ ^

Paola: Can I come and visit you for a little while? Would that be possible?

Sabrina: Let me get a pen here so I don't forget, because I'm not in my little office right now.

Paola: Do you remember anything else that was weird, other than the angel hair and the foil? Do you remember anything, you didn't know what it was at the time, you said 'cause they didn't talk about it. Do you remember the military, coming to see your Grandfather?

Sabrina: Well, at the time I was little and *there was some people coming up there*, and I didn't know what the Hell they were coming up there for. But you know how when you're a kid, you see people coming and going...

But I remember my Grandfather at one time--I was standing by him while these people were talking to him--he didn't speak a lot of English, anyway, they looked like they were Police or something, came up there. And they were talking about, something about coming in the property, and he turned around and he almost got mad and he said, "Well, why don't you just come in from the front, like everybody else?!"

And then I heard something through the fact that he was angry, and you know, they didn't--couldn't come through the front for some reason, like everybody else.

Paola: That's when you remember seeing police or military or somebody?

Sabrina: Yeah, they come over there and, uh, I don't know, *it was like they were telling him not to uh, speak to nobody, you know, like we did something wrong*. And I remember telling my Grandmother, I said, "Did we do something wrong?" and she didn't want to tell me nothing, she just...you know, kids just stay in their place. (*laughter*)

Paola: Yeah, so you got that feeling, as a child, that somebody did something wrong.

Sabrina: Yeah, like, something was wrong. Why did the police come up there? And then you know, like something had to happen? We don't know what; they came over there in a big 'ol truck.

Paola: What kind of truck? Was it military or regular truck?

Sabrina: It almost looked like a Jeep or something, you know.

Paola: A Jeep. OK, so the only thing you remember is that you could touch the foil and the angel hair. But when I come and see you, see if you can remember any other things, because you're telling me a lot when you tell me that your Grandfather was angry. You're telling a lot when you tell me what feelings you had, when you were a child.

As Paola had just noted, our third witness was cross-checking everything we had heard or vaguely suspected from the earlier testimonies of Reme Baca and of Jose Padilla, about the "spider web" and the memory metal. Evidently, there was more of it than even Jose had known at the time. Paola thought that was the end of the interview. But Sabrina was about to add another recollection that extended the puzzle even more.

Sabrina: One more thing I remember is that I had touched that, you know… my Uncle *had given me a piece of this metal that he'd found at the time*, and I had that little piece for a long time, but my kids lost it around here. I had two grandkids. They were fascinated with it and they misplaced it. So, I could have showed you the little piece.

Paola: Oh my God! That is so important Sabrina! That's important if you could ever find it!

Sabrina: It was a piece that broke off and it was in a shape, like I would say a shape of a pyramid, but it was broke. We had that laying around here for the longest time, but my kids got fascinated and they started showing it to their friends. But then nobody believed that stuff, and they ended up losing the damn thing, and I don't know where it is now. I wish I could've kept that so I could show it to you…

Paola: Yeah, but the kids, would the kids know where it is?

Sabrina: No, we looked for it. I looked for it. I didn't start looking for it until my Uncle started telling me about all this (*note: our investigation*), and then by that time we couldn't find it anymore. But he had so many pieces before he moved where he's at now…he had so many pieces like that, laying around the house, that he found when these people (*note: the Army retrieval team*) come over there, and they threw--he was watching them--and they threw a lot of stuff at random away, so they wouldn't have to pick it up, and that's how he came about, trying to get some of those pieces.

Paola: Who was it? Your Grandpa?

Sabrina: No, my Uncle. My Uncle Joe. (*Jose Padilla*)

Paola: Oh, so, your Uncle Joe was the one that had a lot of that stuff.

Sabrina: Yeah, he had a lot of those pieces that...See, when they (*the soldiers*) came over, and they was cleaning up, and taking these things away that they didn't want us to have on the property when the crash happened, or whatever...in the evening time they would get tired of sitting over there, so they would go out for a while.

I guess they go out in town over there, because a lot of people used to like to go out there, and get drunk at that Owl Bar down there: that was where everybody used to go and drink. You could find everybody hangs out over there. I remember I used to like to go over there too at Halloween, because they used to give me a bunch of candy and stuff. I wouldn't get that much candy when going through one house to another and all that, 'cause things were so far from each other. So, I used to like to go. They'd give me whole bags of candy from there, when it was Halloween. That was a long time ago.

Some of those guys that had (been) over there cleaning up and um, my Uncle was watching them, and before they would leave to go to the Owl Bar, they had to clean up, and you know...a lot of those pieces of metal that was out there, they just threw it away over to the side and threw dirt on top of it.

They were trying to do a clean-up job, but they couldn't do it fast. It took them a while. And he went out there a couple of times and collected little pieces and stuff.

I had a real good piece of that metal but, like I said, my kids discarded it over here. I was very disappointed. I wish I could have had that, I'd have gave it to you.

Paola: No, I know...Well, it's just totally amazing that you remember all this, because it's very important: we need people to verify Uncle Joe, so you're really helping us.

Sabrina: Yeah, well I want to tell you this. *If anybody came across that Angel hair, be careful with it!* That's not good for kids to be handling, because I was a kid and handling that. That stuff, you know, it cuts your hands. It's not safe to touch.

Paola: So, what you're saying is that it's irritating. It doesn't make you bleed, but it almost burns your hands?

Sabrina: And I suggest, if anybody finds that and probably don't know what it is, don't touch it. Leave it alone! *Don't do that!*

Chapter Sixteen - Global Patterns and a Surprise Third Witness

Paola: When you say "cut your hands", though you mean it irritated it, 'cause there was no blood when it went in your hands, right?

Sabrina: *It was no blood, but it was, the same time, it was a bunch of pins poking you.*

Paola: Pins poking you; pins. OK.

Sabrina: Yeah, it was scary. I was, everywhere, where your hand touched that thing.

Paola: OK, Sabrina, thank you so much for talking to me. I really am grateful to you.

Sabrina: I remember they had another bomb that they threw back in, oh gosh, it was terrible, you know. I don't think they documented that, but I remember that bomb. When it went out, it just as quick as it happened, it was big, and this was in the evening time.

Paola: I wish I could get the dates. You don't remember if it was Christmas, winter, summer...?

Sabrina: Gosh, that was some time ago. We really had some cold winters up there too, 'cause we didn't have no gas in the house, or anything... We'd get our wood in from the forest, that wasn't far away from there, we'd go deep in the forest and get pieces of wood and bring them in for the winter time.

This is something that happened and a lot of people don't believe us you know, and it's gonna stay in my life. I talk to my kids about it all the time 'cause I want them to know that this is something that really occurred. I'll never forget about it.

OK, well, I'm glad there's people like yourself. And one more thing I forgot to tell you, this is all I can remember right at this point: *That piece of metal, you know, that piece of metal would stay cold on a hot day.*

Paola: That's very important, you remember that it would stay cold on a hot day?

Sabrina: Yes, so I remember when I would get the piece of metal it was, like summer time, in New Mexico it's hot over there. When it gets hot, it gets hot! When it gets cold, it snows, and it's ooh! You gotta wear all your clothes that they had; you know. But uh, I remember when it was real hot, *you could put that piece of metal onto your face and it would be cold.*

Paola: So, you had that much of it?

Sabrina: Yeah, I had that piece, and you know, like I said, my Grandkids lost it here around the house. But you know the thing about it, it was the funniest thing, it was all like a joke, almost. I don't care how tight: "I'm gonna fold this stuff real, real tight and it's gonna stay this way..." And I'll be darned, *it just flipped right back out, like nothing, it had never been folded before.* It was the strangest thing. I'll never... I've never seen or came across anything like that my whole life, since then.

∧ ∧ ∧

Sabrina's first interview was precious. It completed what we'd learned from Jose, and from Paola's interviews of Reme years before that. They kept talking about the angel hair, which they recalled clearly, but they thought it was just filler material between the panels, or insulation. That observation made sense to them, but Sabrina was adding details that contradicted it, adding another level of sophistication to the devices. And it covered the next period when New Mexico, after the war and the atom bomb, was slowly emerging into modern America.

What our third witness told Paola also confirms that Security services were still busy, searching for "something," years after the craft itself had been retrieved—provoking the angry scene at the home of Faustino Padilla when they sneaked in through the back door. They were not from the local police or even the State Police, which wouldn't arrive in a "big Jeep" and would hardly behave this way, casually invading a peaceful citizen's residence.

An expert we consulted about those observations was struck by Sabrina's report that the foil was conducting heat the way it did: it felt cool against her face on a hot New Mexico day, suggesting that the material controlled heat flow throughout its surface. The "pins and needles" impression on her hands also needed a physical explanation. One conjecture is that the "angel hair" consisted of some kind of nanofibers that broke during handling and punctured her skin on a microscopic scale: "This is actually common among people working with fiberglass and carbon fiber," he observed. "It sounds like the angel hair fibers were very small in diameter—not large enough to be seen with the naked eye."

Except for the fact that nobody was manufacturing that kind of fiber, back in 1945. When we analyzed what we had learned from Sabrina, we

Chapter Sixteen - Global Patterns and a Surprise Third Witness

realized there were gaps in our understanding of what she described, gaps in the timing, and subtle contradictions with what Jose knew, possibly explained by the fact that he'd left New Mexico when he was still a teenager while Faustino and Sabrina continued to live in the area. It was time to call her back, nail down the facts that proved the validity of the testimony, and go through the key physical evidence in sequence.

^ ^ ^

CHAPTER SEVENTEEN

ANOTHER ROADSIDE ATTRACTION

In 1977 the MacMillan Publishing Company of New York released an English version of a remarkable Soviet story, written by the celebrated team of Arkady and Boris Strugatsky, simply entitled *Roadside Picnic*. Considered among the best of Russian science fiction, the Strugatsky brothers had placed their very simple story in the quiet Canadian town of Harmont where an extraordinary event has happened, leaving physical objects discarded across the countryside.

In his introduction to the American edition, Theodore Sturgeon describes it as "a brief visit from extraterrestrials," who leave behind them something like litter after a roadside picnic. There are no Aliens in the story. They are long gone but the litter is very real:

"The nature of these discards, products of an utterly alien technology, defies most earthly logic, to say nothing of earthy analytical science, and their potential is limitless. Warp these potentials into all-too-human goals, the quest for pure knowledge for its own sake, the search for new devices, new techniques, to achieve new heights in human well-being; the striving for profit, with its associated competitiveness; and the ravenous thirst for new and more terrible weapons—and you have the framework of this amazing short novel."

We follow the rough experiences of Redrick Schuhart, a young laboratory assistant at a local Institute, who goes illegally into the forbidden Zone, the "Plague Quarter" where the Aliens once landed. He is a trafficker, a stalker, often violating Security, secretly retrieving extraordinary contraptions, dangerous devices the Visitors have discarded: Some release energy, some alter the flow of time, some attach themselves to human bodies and destroy them, and some, like the Golden Ball, are simply beyond human ability to characterize them and beyond human imagination to study them because they take control of your miserable mind and engulf you in eternal bliss before you can think about it.

As part of his job, Redrick sometimes takes eminent academic scientists there on official tours, marveling at their face always "calm and clear," with no idea of the dangers they would encounter if he wasn't guiding them: "They're all like that, the eggheads, the most important thing for them is to find a name for things. Until he had come up with a name, he was too pathetic to look at. But now that he had some label like *graviconcentrate*, he thought that he understood everything and life was a breeze."

As we completed the cycle of our research at the San Antonio site, we didn't have any new theory about *graviconcentrates*, or a neat display of devices retrieved for us by stalkers like Redrich Schuhart, with his wonderful inventory of *Full Empties, Death Lamps, So-So's or Witches' Jelly*. Instead, we simply knew of four unusual physical devices associated with the San Antonito crash, a real event for which we had independent descriptions from at least two, and sometimes three witnesses.

There was the bracket, of course, recovered from the inside of the craft where it was pinned to the wall; there was the memory foil; there were the fibers bundles; and finally there were metallic items.

We quickly agreed the bracket was not unusual in composition although we disagreed on its purpose. My friends thought it might be part of the original equipment aboard the craft, while I argued it might have been brought in by the soldiers to facilitate their work of recovery inside the vehicle. The metal items were similar to objects we had collected at other places (notably at the later crash site near Datil) which we had already analyzed as ordinary aluminum, although we had no definitive explanation for their presence at the site.

Chapter Seventeen - Another Roadside Attraction

That left the memory foil and the fibers, for which nobody could account. Now our purpose was simply to improve our understanding of some, or all, of these devices.

^ ^ ^

Paola's second interview with Sabrina took place in December 2020, and it began unexpectedly with a shocking revelation that something or someone was interfering with our investigation: Sabrina was surprised that Paola was calling her, because she'd heard on the phone from "someone" *that we wouldn't come to record her detailed testimony after all.*

Naturally, Paola was somewhat shocked so she asked, "How soon after I called you and introduced myself did you get the call? How many days would you say?"

Sabrina: That was *the next day*.
Paola: The next day?! OK...
Sabrina: That's why I believed the phone call was coming from somebody that you were affiliated with, because they said, "Oh yeah, we know her. But she's not gonna to be able to come. We're gonna come and take care of everything."

And I said, "OK!" And that was the end of everything, right there, just like that! It was like, (*chuckles*) "Bye, hi and bye! And never see it!" I thought, "Oh I guess she couldn't travel, maybe." You know, I believed it 'cause they said "Paola", and I don't know a lot of people named Paola, to be honest to God. I don't know anyone but you with that name.
Paola: Well, tell me, what were they were going to take care of? Did they mention anything about this case?
Sabrina: No, they just said, "Oh we're gonna come down there and take care of that." You know, because you weren't able to come. And I said, "OK." Just like that and that was the end of the conversation, you know.
Paola: Did you honestly think I would send somebody without me?
Sabrina: Well, I don't know, because we'd been talking before, but I just thought, "well maybe she got held up and she's gonna have somebody else come out here and talk to me." I thought about it in that way, and I believed that for a minute. I'm glad I mentioned

that, because *I see now that these people aren't who we thought they were.*

Paola: I don't know who those people are Sabrina, and neither does Jacques, because we never mentioned you to anybody.

Sabrina: Yeah, that's strange. It's a strange call altogether. But, right now everything, a lot of strange things are happening by the minute, seems like, my gosh!

Paola: OK, so Sabrina, what did the voice sound like?

Sabrina: It was a guy. And he sounded kind of mature, on the phone. He didn't give me his name, but all he mentioned was, "Oh, I know Paola. She won't be able to come." We both hung up and I went back to doing what I was doing, because I believed that maybe you put him up to calling me, you weren't gonna make the date, what have you... That was right after you called me, the next day, in fact.

Paola: Tell me something. Did he give you a date he was coming?

Sabrina: No, I didn't even think of that. I figured he was coming on the same date that you were coming, because I didn't give no date. I just mentioned, "Oh, well she's coming." (*and he said*) "Oh, I know her, she's not going to be able to come, we're gonna take care of that." And I said, "OK." But that was the end of the conversation, we didn't exchange dates and things and behind it coming right the date after you called me, I figured it was alright, that you knew these people. But I'm glad now that I mentioned it to you 'cause I see now that I don't know them and you don't know them, something really strange. Oh boy!

Now we were facing "...something really strange..." which also happened to be illegal, because it represented a serious breach of Paola's telephone privacy, and a crime. While some organizations on the fringes of the Intelligence community have played such games before, intercepting private communications isn't above the skill level of many hackers these days, nor is it very hard to impersonate someone over the Internet or the phone. It has long been clear that rights to privacy have been silently compromised, decades ago, without much resistance. But the timing was suspect and disquieting to us.

After we argued about it, I told Paola the guy was probably just another jerk from Foggy Bottom and that I wanted to ignore him. We decided that this incident, which demonstrated that "someone" was paying close

Chapter Seventeen - Another Roadside Attraction

attention to the ongoing study, would not push us to resort to encryption or to devious methods of hiding our travel schedules or phone conversations. If the intruder was intent on tracking us, he would find other ways around any technical tricks; and if he was part of the government, he would have legal access to any numbers or encryption keys.

We were doing nothing wrong, so the conversation continued as if nothing had happened, and Paola went over every detail again.

^ ^ ^

Paola: Well, the main reason I called is, I just wanted to just ask a few more questions. You're important to us Sabrina, *'cause you saw all four types of metals*, beside Jose and Faustino. When did you arrive at the Padilla Ranch? What year?

Sabrina: Oh, that was back in '53.

Paola: And how old were you when you started living with your Grandpa Faustino?

Sabrina: I was like, two months old.

Paola: Two months old in '53, ok. And you began to understand those things when you were like six, seven or eight-years-old, right? The early 1960s?

Sabrina: *I walked up there to the crash site and saw it for myself then.*

Paola: Well, do you realize that's almost... it had happened in '45, so it was like fifteen years later. Do you realize that? Fifteen years after the crash!

Sabrina: That's right. That's right!

Paola: And it still looked bad?

Sabrina: It looked terrible. In fact, I would have walked deeper into there, but it just gave me ...you know how you get this scary feeling over you, like "don't go there?" That's what happened to me. It was a big area to me at that time, the way I can describe it to you. It was scorched black, *scorched*! I don't know if you're from California, but we have a lot of fires out here. And when something gets scorched black, like a whole hill that's burnt up, it gets real black ....that's the way that looked. But what I'll never forget, I saw a tree that used to be there, and it burned up so quick that the skeleton to the tree was standing there, dead.

Paola: But the problem that I'm having is, what are you comparing it to? When do you remember the tree was normal? Like what age do you remember it was normal?

Sabrina: Actually I never really paid too much attention to the tree before, 'cause there was a lot of shrubbery around there at that time. Just wild, you know, plants. They used to call it *Yerba Buena*. It used to grow very wild around there and it just grew all over the place. There was a lot of shrubbery and mesquite.

Paola: Sabrina, you're not going to believe it: They've changed the site so many times since I've been there! And you said it was scorched all over, and then you saw some kind of "indentation that was deep." But the excuse they're using now, is that they're building up a dam so the water won't flood.

Sabrina: Ohhh! Oh, that's weird. I don't know what they're doing now. I'm surprised that at this point, they haven't built some houses on top of there.

Paola: No, they haven't built houses, but I'm trying to figure out how much dirt they put on what you saw. Did you only see it once?

Sabrina: I think I've only seen it maybe twice but, you know, that was a real long time ago Paola, to be honest with you. It looked the same way when I went back. But then after that, *my Uncle told me there was a real bad storm came through, (and) there was like a tidal wave, that came through with a lot of water. It might've changed some things. A lot of dirt might have went over, 'cause they had a real bad catastrophe there that killed a lot of animals.* There was a bad stench there for a real long time.

Paola: When you're looking, how far away were you from the actual indentation?

Sabrina: I wasn't that far from it 'cause I'd walked so far already, but I got scared so I came back home.

Paola: But how far away were you from it? Is there some way you could show me?

Sabrina: Maybe fifty yards, as I say, it was like from across the street from my house south across from me, that's about how far I was from it… fifty yards. It was just like a bunch of dirt. Like a little hill. It was like a little tiny hill, like…I don't know if you ever seen something happen like an avalanche or something, when something real hard is rolling down and it makes an indentation and goes into the ground,

Chapter Seventeen – Another Roadside Attraction

you know. You almost have to dig it out, or whatever, to see what made that. Sometimes rocks make things like that.

Paola: I'm picturing what you are saying. So, it wasn't really that deep of an indentation.

Sabrina: Yeah, 'cause you could see where something fell and slid to make that indention, like that it fell and it slid.

Paola: But you also saw the slid part? The part where it slid?

Sabrina: Yeah. It was just kind of, it was just odd, you know, 'cause everything was burnt up and it was scorched.

Paola: But you said "slid." So, you would have to see something that let you believe it slid.

Sabrina: Yeah, it disappeared that way, that "something," in the way it fell. It didn't fall right in that spot. Before it stopped, it slid a little bit, you know.

Paola: That's perfect, Sabrina. I thank you because…you're telling me what Reme told me, because when he described it, he said it slid and made an L shape turn on the ground.

Sabrina: Yeah, so I hope I was able to help you with that.

Paola: Well that helps, and then I can bring some photos with me to show you how they've made a mess of it over the years.

^ ^ ^

That exchange was important, and it is reported here in detail, because it filled in the history of the crash site between the time when the kids were there and the time when Paola started documenting it, years before I became involved. It also established that any samples casually thrown away by the soldiers in a crevice in the earthen dam would be unreachable now, under some twenty feet of tightly-packed dirt, with a pond on top of it. But this also brought up the subsequent visits by the military, coming back to the site years after the main event, for reasons we couldn't clearly understand. Sabrina's testimony was precious to fill the gaps.

When the conversation continued, Sabrina was asked about the Army's operations at the site and she commented: "coming over there and stuff, I remember when they were fixing my Uncle's fence out there…I don't know why on earth they kept coming through the back over there, where the arroyos were. They wouldn't come through the front, like everybody else, you know. We had this long, long driveway for our property. They wouldn't come there like everybody else. They wanted to come through the back of the property. You know we kept the cows ...

Paola: But who's "they"? One guy at a time or a bunch?
Sabrina: Well, when that Military came up there and made that opening, it was like a big arroyo there that went under a bridge. They, uh, they made it so anybody could drive through there. And I remember, like I said, that same cousin of mine, he used to drink a lot and he'd get drunk and we'd get in the truck and he took us on a ride down there and back. We weren't supposed to be driving through there but we... (*chuckles*).
Paola: How do you know you weren't supposed to be driving through there?
Sabrina: 'Cause everybody just took the main road that was in front of our house, that went all the way down, you know, to someone else's house. 'Cause see, we didn't have blacktop pavement over there, this was just dirt road. But that right there, we didn't ever use that. That's something that they came and they made.
Paola: Did your Grandfather say the Military made that road?
Sabrina: Well yeah, they are the only ones that could've made it there. Nobody else would make anything there. See, there was nothing but arroyo come through there.

∧ ∧ ∧

That sense of the continuing presence of the military is something that comes back in all testimonies from our three witnesses. While it isn't particularly concerning, it shows that the area was being watched continuously and that access to the isolated site was being maintained. The conversation also built up the background upon which Paola could go over the information we already had about the whereabouts of our famous "bracket," its chain of custody, and validate what we knew from Reme Baca and from Mr. Padilla.

Paola asked: Let's talk about the "bracket." You said that you saw the boys bury that "bracket" piece. You told me that. Can you describe that? Did you come out of your house, and you saw them do something?
Sabrina: No, *they had already buried it up under the house.* It was right in next to the house, in these rooms, we used to call them *Viroca* (*Note:*

Chapter Seventeen - Another Roadside Attraction

An annex?), big rooms, I used to sleep in there sometimes. He didn't tell me that they were burying that under there.

I just knew that they had buried something under there, 'cause I seen the dirt where they had been, you know, digged up all that dirt and put it back, you know it was right near an old barrel we used to keep tar. You know that black tar? We used to use that to do roofing on the house. It was not far from that. I didn't ask too many questions. I wasn't too concerned about it anyway. I was minding my own little nine-year-old business...(*laughter*) I just thought, well maybe they're doing something there, 'cause, we actually did a lot of things in that back room. We used to target shoot. We used to have a little, tiny window and there was like a mountain right far from it, kind of next to where the cows used to graze. We had an old silver milk tin, you know how they used to put milk in those, almost like the size of a barrel, and we used to shoot at that thing all the time.

Paola: You know, we're here today. But what I'm trying to do is get *all the story, before you knew what that was.* When you said the boys "burying it"... I haven't talked to Jose about what you said, so this makes it even better, because Jose doesn't know what you said, and you don't know what he said. You're perfect for us Sabrina, because you don't know (*those statements*). But I'm wondering if you remember seeing them: Did you actually see them bury the things, or do you just know they did?

Sabrina: OK, I know that that particular year, we'd went deer hunting. It was a real good year, 'cause we caught two deers and that was a lot of meat for us and it carried us practically through the whole year. We even brought some of that meat to California to my Aunt and other people in the family, 'cause we ate deer--deer meat I should say. And uh, that particular time, what I could tell you, 'cause I was a kid, is they either put it (*Note: the metallic bracket*) under the house or took it out from under the house. One of those two things happened. My Uncle must have put that back then, all them years; it probably stayed there, but it didn't get taken out 'til the time I saw that they digged up under the house.

Paola: OK, because he did take it to California, he did, thank God, because then the house burnt down.

Sabrina: Yeah, the house burnt down. It was there for a very long time, you know. I remember I didn't like to go play back there too much. We had uh, a big ol' can there, not a can, what you call them, "barrels", a barrel what we kept tar in. And then one time, my cousin that I used to live in the house with at the time, he killed a snake and threw it back there. And Paola, I don't like snakes, ok?

Paola: (*Laughter*) I don't either, I don't either...Oh, my God!

Sabrina: I would take my pistol out and shoot the son of a bitch, you hear me?

Paola: I can see it all, I can see it, I'm terrified of snakes. Did you ever see the piece (*the "bracket"*) in your life? Did Jose ever show you the actual piece he took?

Sabrina: Oh, of course, I seen it many times! I seen it, uh, when he was out here, in California, he took some pictures of it, you know, a real long time. It was basically the same material as the little piece of metal that I had, that unfortunately we lost. What I could describe to you, it was very light material. It's light as a feather. And I think if somebody else would get a hold of that, they would probably even toss it, and not even realize, you know, where it really came from.

Paola: Well, my question--we're going to ask you some more when we see you--is this: *I know what your Uncle Jose took, but I don't know what your Grandfather Faustino took.* This is the problem I'm finding out from you, that *your Grandfather took materials also.* I didn't know that, and neither did your Uncle.

Sabrina: Well, you know my Grandfather used to get out on the horse, and go out quite a bit. I seen him go out at least, maybe two, three times, and then after a while he couldn't ride no more 'cause he had stomach cancer, and he wasn't able to ride anymore. I was the only one who used to ride the horse around the property there. My other friends that lived up in the hill up there, the Reeves, I was jealous of them because one of their kids had horses, and I used to think that at that time, that was a big deal to have your own horse. They had a pony and these other two horses. I used to go over there and ride their pony sometime. I stopped riding him after that time when I got on him and we went, we were coming on my way to my house and a snake scared it and it threw me off and I never did ride it no more.

Paola: Oh Lord, yeah, you must've seen a few snakes, I imagine. OK, well, your Grandfather, *not your Uncle*, did a lot of things that your Uncle Jose didn't know, because nobody knew your Grandfather went back there and got stuff. And yet your Grandfather knew that was no weather balloon. He knew that!

Sabrina: Of course. Of course. You know after my Uncle Jose left, it was just us alone there, you know. It was just us alone. It was just me…

Paola: Yes, when Jose left to join the military. He was like 15 or so.

(Note: Mr. Padilla was born in 1936, so this would have been about 1951 or 1952.)

^ ^ ^

The next item we had wanted to verify with Sabrina was what Mr. Padilla had called "the angel hair" in previous conversations. To me, working from his description, it would have been described more accurately as fiber bundles, which formed clusters, as opposed to angel hair, which extends into individual strands.

Paola: So, the second thing you told me you saw after the bracket was the angel hair. You said it was like human hair; and it was very silky.

Sabrina: Very silky, shiny, *and at night it would glow.*

Paola: And what color did you say it was again?

Sabrina: It was like a purplish-pinkish looking color. Very light, but it was pretty, to me.

Paola: Was it on a strip? Did it hang down on a strip?

Sabrina: No, it was just a big ol' glob of it.

Paola: It was a big "glob" of it? It was a glob of it…

This was new. Mr. Padilla had not mentioned anything of the sort. And the color, bordering on the ultra-violet, was very interesting.

Sabrina went on: There was no end or beginning to it. It was just a piece of hair and it was so long, it was like about, uh… How long would I say it was? Maybe about, uh, maybe about the size of a ruler. About 12 inches.

Paola: But you saw more than one of those, right? You saw more than one angel hair clump?

Sabrina: Oh, absolutely. There was a whole bunch of it. I had it in a plastic bag and we used to basically just keep it in there 'cause that thing, it was irritating. You couldn't handle it too much. It was very irritating, you know. Even if you got one hair, one little strand would bite you. You didn't need all of that hair to attack you, just one little strand. That's why we couldn't play with it. (Note)

Paola: (*Laughter*) It would "attack" you... It was like electricity, you mean?

Sabrina: Yeah, yes, that's how the way it carried on. That's exactly right. I don't care how you touch it, you know, it was just irritating to the hand, I don't care any part of the body. So, imagine somebody would've put that on their head, or something like that, oh God!

Paola: Well Sabrina, tell me though: *nobody ever talked about this before*, like, your Grandfather Faustino didn't say, "Oh I found this outside..." Your Uncle Jose would never say, he found this over there. So, you just saw strange stuff in the house, but you didn't know where it belonged, or where it came from?

Sabrina: Yeah, I saw strange stuff all the time! (*Laughs*)

Paola: 'Cause your Grandfather went to the crash site, the day after actually. *And he went inside the craft*. And I'm wondering, where did he get those clumps from? Did he go pull them off from inside the craft? Where did he get 'em? And now you're saying, "Oh he went back, a year later (*note: to the site*) and he happened to find stuff."

Sabrina: He might've had those things there all the time, 'cause after my Uncle Jose went in there, right after him: after he told him, he probably had to go. He just didn't take his word for it. I'm pretty sure he went over there and checked it out himself. Faustino went back to investigate that, and found some things himself and brought it back with him. That's the only way, 'cause when you're a kid, and let's say you take something, you're already scared 'cause you went over there, and took it. The reason I say that my Grandfather probably went back to investigate is one story. And my Uncle Jose had already took off with a piece he found (*the bracket*). You see what I'm saying?

Actually we now know that Jose never told Faustino he had extracted the "bracket." We also know that Faustino went inside the craft first, along with the police officer, leaving the kids outside. Sabrina may not have been aware of the actual sequence.

Note: In a follow-up conversation on February 12, 2021, Sabrina confirmed to me the material was "very strong, *not attached*, pink, silky and smooth."

Chapter Seventeen – Another Roadside Attraction

Paola: Well, the reason this is interesting is that you've had access to all four "metals."

Sabrina: Yeah, that's crazy, but he went out there on his own and got a hold of that stuff and brought it back. 'Cause we also had a bunch of this. It was this real weird hair, they got a whole gunny sack full of it and brought it home.

Paola: You think that once Jose told his Dad the story, Faustino went and looked for more?

Sabrina: He probably went out there and got some stuff and brought it back. We only had one horse so, all of us had rode that horse, even me. And he went out there and picked that stuff up. But I remember my Uncle at one point telling me that when that thing crashed over there, that made a hole on that, um, *craft* or whatever that it was. A lot of that stuff was all over the ground and they just picked it up and put it in a gunny sack. It came from inside the ship – like it was some part of the interior of it or something.

Paola: Yeah, that's the so-called fiber "insulation." I always wondered if we could get some of that from your neighbors. It was all over the place.

Sabrina: Well, to be honest with you, Paola, all those people are probably dead now, it's been so many years. A lot of their families are not even there no more, or their kids don't even have the property no more. I remember we just kept a little bit of it in a plastic bag and I used to play with it all the time. But I would never play with it too long 'cause it would bite you. It was irritating to the hands. You wouldn't dare put it on your head.

Paola: (*chuckles*) No, I know. I love your story, because your story is different from your Uncle Jose's but it completes it.

Sabrina: That's what I think happened. When I talked to my Uncle, he said, "Oh, when that thing blew open, all that hair splattered everywhere. We picked it all up and put it in a gunny sack."

Paola: Yeah, I know, he actually brought Jacques there and he showed him where he got it and everything. I have a movie of his walking there with Jacques and showing him everything. Your Uncle is amazing.

^ ^ ^

After the bracket and the fiber clusters, we had yet to fill the gaps in our knowledge of two other items, namely the "foil" and what had become known as the "pyramid," two items for which we were at pains to guess the possible function. Paola began with the foil.

Paola: You're so good Sabrina, the way you described that stuff. But I'm curious about the foil though because the Military seemed to be really interested in that, so your Grandfather was hiding it. The foil – where did he put that?

Sabrina: He was hiding quite a few things! (*laughter*) I got a chance to play with it once. It was just the strangest thing I ever saw. I thought it was a joke.

Paola: Reme only had a little square of the foil. I'm wondering how come Faustino had long strips of it. Did he go out on his horse and get the rest of it, do you think?

Sabrina: Yeah, he probably did 'cause he used to go out a lot, and I used to catch him when he was coming home, because as a lonely child, I only had a dog to play with. When I didn't see the dog, you know--we used to have this dog my Uncle gave me, and I said, "Oh they went up there to the mountains again." And I remember when they came back, the dog and the horse...

Paola: OK, because we don't know how your Grandfather got those long strips of foil, we can't just imagine it. But where did you get it when you got a chance to play with it? Did he give it to you?

Sabrina: He let me hold it and play with it for a little while and then, you know I got tired of it, just like you do with anything else. You got to remember I was a kid, you know. I remember it wasn't that big. It was like maybe three or four strips. It looked like foil.

Paola: Was it in separate strips?

Sabrina: Yes, they were separate strips.

Paola: And how wide were they?

Sabrina: Oh, maybe an inch and a half. Like the size of a ruler, you know?

Paola: Yeah, that's good, that's good, the size of a ruler.

Sabrina: And the only thing I remember about that there was like maybe three or four of them strips and I don't--Paola, this is crazy...I don't care what you did to that foil–if you balled it up, like you're throwing a piece of paper in the trash, and you know how you ball it up? Or if you get it and fold it...It was very light. It was light as a feather. And if you even got it and folded it up like, you know how you do that Origami, how you fold things up?

Chapter Seventeen - Another Roadside Attraction

Paola: Yeah…

Sabrina: I don't care how you folded this piece of foil up, it would straighten back out flat.

Paola: OK, and you were nine years old, so we're talking like, um, 1960 or something, right?

Sabrina: It was in the '60s 'cause I remember, President Johnson was President then, and me and my cousin Genevieve, that I used to go to school with, we used to sit by the butane tank, at the school in San Antonio. They had a big butane tank over there. Why on earth they used to allow us to…? We used to sneak away and go sit under the butane tank, and eat our sandwiches for lunch and we used to talk about President Johnson. And she said—you gotta remember the conversation between nine-year-olds, you know--"He's sick all the time. My mother says he's got cancer in his stomach." Just things like that.

In analyzing the transcript from that conversation, we had to marvel again at Sabrina's fine recall of details that were like pointers to the accuracy of the testimony, in the context of the elements we previously gathered and cross-indexed. So we kept going on firm ground.

Paola: Now, you had that stuff, or Faustino had that stuff since 1945. It would have to be about 1964 (*when you saw this*). 'Cause if (Lyndon) Johnson was president, it was about 1964. So, I don't understand, what did he end up doing with it?

Sabrina: Well, I don't know, I think, um, he ended up giving it to one of his friends. But this is the story on that: it ended coming back to us, in fact, 'cause I don't know if it's documented, but we had a bad tornado come through there. I don't know how we survived that tornado. It was real bad. And it knocked our windmill down over what we used to call the *Viroca*. It knocked the windmill down. It broke it like a toothpick from the bottom. And it laid over on top of the building, which we were able to recover later and get it off of there. Anyway, we had to fix the windmill because that was our source of getting water. We didn't have no…we barely had indoor plumbing.

So anyway, we were trying to repair the windmill, which took a long time. We had to rebuild the frame and then the hardest part was to put the big ol' round thing up on top and then they discovered at that time, it needed a certain part and we didn't have it. So, my Uncle had his friend helping him and he told him, "Do you still have some of that, you know, that little foil, or something that I gave you one time?" He said, "Yeah." So, he brought it back over to the house and they put it on the windmill, believe it or not, it fixed that windmill for over, must've been 70 years until they took it down.

Here Paola had to step in, because Reme Baca had claimed that he'd fixed the windmill with the foil. Reme was probably the "friend" in question. While the use of the material to solve the windmill's accident checked out in Sabrina's version, it seems Reme may have exaggerated a bit his role in the actual repair.

Paola: OK, now I'm really confused because it was Reme Baca that said he fixed the windmill with that piece, and Reme only had one piece. Jose saw that piece of metal, too. But I've got to ask him if it was a square piece or if it was a long strip. What do you think? It's the long strip he fixed the windmill with? Is it a long strip or was it just a little square piece?

Sabrina: I don't know if he might've seen that. But I know when my Grandfather took those out and let me play with them, *they were long strips. They were like the size of a ruler*. They were long and the width of the ruler. And they were like, three or four strips, and I played with them 'til I got tired and then he got 'em and put 'em away someplace, and I never saw them no more.

Paola: And he never, of course, said anything, right?

Sabrina: No, we didn't talk about it much...I don't really know what they did with all the pieces. Or maybe they just threw them away.

Paola: You saw more than one piece though, didn't you? So, that's our third kind of metal. The Foil. If you had to guess, how many pieces of these strips did you see, on the table?

Sabrina: Oh, if I had to take a guess, I would say about three. 'Cause I used to play with them. I played with them until I got tired of them.

∧ ∧ ∧

The fourth item we needed to check out from Sabrina's story was the small pyramid, unfortunately misplaced.

Chapter Seventeen - Another Roadside Attraction

Paola: Where did you get the piece that looks like a pyramid, that you put in the CD case, where did that come from? That's your fourth kind of metal. What I'd never understood, was that what was in the CD case was metal, and not a piece of foil.

Sabrina: I think we had that laying around the house. My Uncle had it laying around the house, and he just gave it to me.

Paola: And you don't remember where that pyramid that you had for the longest time, that would turn... How did you know it was special?

Sabrina: Oh, I knew it 'cause when I got to see the part that my Uncle had buried up under the house, *it was the same material*, light as a feather. And you know, it was the strangest thing. That thing was strong! Strong, strong, strong! And it would stay cold, a lot of time.

Paola: Why are you saying it was "strong"?

Sabrina: Well, 'cause my Uncle told me one time he tried to burn it. He tried to put a blow torch to it and nothing happened with it. It wouldn't melt or anything (*chuckles*), and it would stay the same and it stayed cool. A piece of metal, if you try to put any heat to it or anything, it will turn hot and you can't touch it or it will burn you.

^ ^ ^

If we thought that was the end of the items Paola needed to go over with Sabrina, both of us were wrong because yet another anecdote emerged from her memory. While it may not be directly related to the crash or its aftermath, it completes the picture of what New Mexico was like after World War Two, and it brings to life the facts of life in rural America that may be hard to imagine for us now, in the world of abundance of the twenty-first century.

That curious mention of the strange rocks with which the little girl used to play came up as Paola was ready to say goodbye to her.

Paola: I just wanted to get some clarification 'cause *you're the only one that experienced all FOUR pieces*. You had all four things in your life. And what about the strange rocks you picked up?

Sabrina: I don't think those rocks came from that, 'cause they weren't there (*note: at the crash site*). They were in the cemetery. I remember when I brought them home, because when I was a kid, I used to go out there and play in the cemetery. It was kind of morbid. You know,

I was a lonely child. I brought the rocks home. My Grandmother got mad ...

I liked the rocks 'cause they would glow in the night. I don't know why they would glow or what, it was something about that. I got about, maybe ten or twenty of them little rocks. I could carry 'em in my hand and they would glow. When I brought 'em home she said, "Why did you do that?" And I said, "Why, I found these rocks..." Even though they were from the cemetery. And she said, "Tonight the *Muertos* are gonna come and get you for those rocks!"

Paola: Yes, the dead people...

Sabrina: Exactly! Oh! And you know you're telling that to an eight-year-old, right? The next day, I got up and took the rocks back and threw them back over there. *(laughter)* I didn't want "them" to come for me in the night for that stuff. It was serious, 'cause it was like desecration or something. So, I went and took the rocks back.

Paola: But Sabrina, rocks don't glow in the dark, so what were they?

Sabrina: I have no idea, 'cause you gotta remember I was like eight or nine-years-old. They were white. I know that the rocks were white. And they had a little bit of silver in them. Like little indentions of silver. But you know, and that's what got my attention, I used to like 'em 'cause they would glow in the night.

Reading that transcript, I had to reflect on the dangers to which the child was exposed. If those rocks were ordinary stones that had been irradiated by the 1945 explosion, that means they were still a source of harmful radiation fifteen years later! Sabrina's Grandmother may have saved her life when she ordered her to take them back to the cemetery.

Paola: It doesn't say how those rocks got in the cemetery, 'cause somebody put 'em there.

Sabrina: It's a possibility, they threw them over thinking...well, you know we had a lot of these young soldiers then. So, they probably said, "Well let's just throw these up in there!" They won't think, you know, anything from a cemetery. They probably threw them up in there. It's possible that those might've came from that, 'cause those were strange rocks. My Grandmother scared me so bad about the rocks so I took 'em back. *(laughter)* Threw 'em back in there! I still worry about that, that tormenting, that's an awful thing to do to a

Chapter Seventeen - Another Roadside Attraction

child. It bothers me. I'm 60-something-years-old, you know and that still bothers me.

^ ^ ^

Paola: Let me just ask just a few more questions. Did your Uncle ever say anything about the main event? Did he ever talk to you about it, at that age?

Sabrina: At that time, we all knew that something had happened on the property. To me, it didn't concern me so much, I just knew that there was a lot of people going in and out. A lot of police was coming over, people in uniforms. And I would wonder, what did we do wrong? Why are they coming? I still didn't understand at that point what was going on and why they were coming in and checking so much.

Paola: You said "checking," Sabrina. What does checking mean?

Sabrina: They would come over periodically and see my Grandfather and talk to him for a while. It's like they just checked in with him periodically and asked him questions. Sometimes they would stay ten minutes, sometimes they would stay five minutes and leave. And just look around. I don't know what they was looking for. They never really came in the house. I just saw them outside.

Paola: And they would look around your property?

Sabrina: Yeah, they would just like look around over their shoulder and just, you know when you go to someone's house, some people have a tendency, whatever stands out at you...you just keep looking. For some reason that's what was going on, but they wouldn't stay very long. It was like they had to make this stop and then leave, you know. And that happened quite a few times.

Paola: Do you remember the year?

Sabrina: I would have to say, that would had been in the sixties.

Paola: The other thing, really quickly-- *Do you remember what branch of the Military came to see you?* You said "policemen." So that would be probably Highway Patrol. Or was it Army, Air force? Who were the actual military, do you remember? (...)

Sabrina: They were looking--to me they were like police, and I was just kind of puzzled at the time, because you know, I wasn't really too concerned what was going on with that over there, being that it happened like an "accident" to me. But I got so used to them coming around there, when they started coming like that, sometimes they would only stay five or ten minutes and leave us, like they were checking on something. I don't know, it was almost like they came by just to check and say hello. And then they would talk to my Grandfather of course, and I would just go on about whatever I was doing at the time.

Paola: But you knew there was an "accident" that happened in a certain place, you had that feeling?

Sabrina: Of course! It was always an eerie feeling. I remember that's the part that when I went up there, and even after all them years, that place looked terrible. I didn't think about it, that maybe somebody died over there, nothing like that.

Paola: OK, that's what I was going to ask you. What were your feelings when you walked?

Sabrina: It gave me a frightening feeling actually, very eerie. You know, "I shouldn't be there," and I was scared, I'm not gonna lie to you, *I was scared...*

Paola: ...scared because you saw the military, or because your Grandfather said something?

Sabrina: Oh no. I was scared of the way the place looked...the crash site. I wasn't as scared of the military or the Police. What I was scared was the way that place looked over there... the way it just...*something wasn't right, you know.* It was just unexplained, wasn't right. You could just see it on the ground that something had occurred there, that wasn't there before, you know, 'cause I remember seeing it when I was nine-years-old when I seen it all scorched up and everything. And when you think about it really, you know my poor Grandparents, they didn't watch me all the time, like they should've. I could've got bit by a rattlesnake or something out there, all by myself, you know that wasn't a place to be wandering, 'cause there is snakes over there.

Paola: Oh, don't keep reminding me! I've seen them every time I'm there. It's been a lot of courage for me. And I've been going there for nine years now, and your Uncle Jose killed one right in front of me once. So well, you know Sabrina, that's all I'll ask you.

∧ ∧ ∧

Chapter Seventeen - Another Roadside Attraction

Now we had as full a picture as we could get about the recovered materials, short of actually getting hold of them and bringing them into the lab as we would have loved to do. The only material evidence we did have in our custody was the bracket itself: neither the fibers nor the foil or the pyramid were forthcoming, and we had given up on the idea to recover anything more from the site, newly piled up with large amounts of dirt, and a new pond on top of it.

When I first held the bracket in my hands, I entertained the idea that it had been set up by the soldiers to assist in the recovery and cleanup of the crashed vehicle, perhaps as part of an improvised electrical energy supply, but a senior NASA engineer I consulted, who followed the research Paola and I conducted at the site, disagreed, observing that the bracket...

> ...was reported to be part of the circular copper colored plate that was designed-in to the vehicle. That means that whatever its function was, it was part of the essential design of the vehicle, and the fact that the bracket dimensions match those of the copper plate means that it was designed with knowledge of the copper plate design.
>
> The only way that could happen is if the bracket was designed by whomever operated or otherwise had access to the vehicle.
>
> But the vehicle design was way beyond the scientific and technological knowledge of the time and place. Why would the vehicle operators use a 1945 technology casting for an essential part of their back-from-the-future vehicle? My conjecture is that it was a field repair of a broken or otherwise nonfunctioning original piece.

Chemical analysis has now confirmed that the bracket was not a high-technology item but was consistent with contemporary knowledge of metallic alloys.

The fibers are more interesting and a continuing puzzle, since they exhibit unusual luminous properties. The foil is interesting as well, but it was only a few years later that materials with the property to preserve their shape became known in American industry. The suggestion must be considered that recovered material from UFOs may actually have inspired, or accelerated, classified projects to implement such properties, but the idea of shape preservation was not beyond American or German engineering in the 1950s, and perhaps before. What we simply fail to understand is the actual propulsion of the craft, for which we have no clear data. We don't even understand what caused the collision with the tower that led to the actual crash.

^ ^ ^

In Sabrina's extremely valuable testimony, which Paola took great pains in reviewing and verifying with her as we just saw, the ability of the material to control temperature is the most interesting feature: She clearly stated (and repeated when I met with her) that in hot weather, the metal would not get hotter but would remain cool on the skin, which the child actually enjoyed as she innocently played with it...

That feature alone may lead some researchers to conclude that the object that crashed was of unusual manufacture; and we have seen that comparisons with the properties of other vehicles, like those seen and so painstakingly analyzed by multiple experts at Socorro and Valensole, do reinforce the extraterrestrial hypothesis. But what does "extraterrestrial" mean, in our current, limited understanding of the structure of our cosmic environment?

The question has come up in other remarkable episodes, inspiring the authors of a recent scientific article to note: "It should be strongly emphasized that proving that something is extraterrestrial is extremely difficult, *even if one had a craft in hand* (my emphasis – JV). One might imagine that the presence of unidentifiable, or incomprehensible, technology would constitute potential evidence. However, it would not rule out the fact that it could have been created by someone on Earth." (Note)

Note: See "Estimating Flight Characteristics of Anomalous Unidentified Aerial Vehicles" by Kevin H. Knuth, Department of Physics, SUNY-Albany, with Peter A. Reali and Robert M. Powell, Scientific Coalition for UAP Studies (SCU) published in *Entropy* 2019, *21*(10), 939. https://doi.org/10.3390/e21100939. Special Issue: *The 39th International Workshop on Bayesian Inference and Maximum Entropy Methods in Science and Engineering*, September 25, 2019.

Chapter Seventeen - Another Roadside Attraction

Fig. 30: Discussing Sabrina's testimony: Socorro, February 2021.

The image that will always haunt me, when I think of Sabrina's testimony, is that of a lonely (but curious, and very sensitive) little girl riding her horse to the scary, carbonized remains of the site where the strange object had crashed many years before, and pursuing her excursion to the cemetery to pick up those strange stones that glowed in the dark... fifteen years after Fermi and Oppenheimer, "the destroyers of worlds," had secretly triggered the first atom bomb.

^ ^ ^

The most puzzling element in this whole story, even more exciting than the hypothetical propulsion modes of the avocado, is the nature of the little men seen by Jose and Reme near Ground Zero, by Lonnie Zamora in Socorro, and by Maurice Masse in Valensole. If they are the denizens of some other planetary system, why are they so well adapted to the earth environment, and why do they breathe our air without suffocating?

On the other hand, if they are humanoid creatures from an undiscovered race on Earth, where is their home port? Could they be artificial humans, a sort of biological robot built by an adversary of the US?

One man was approached by the CIA to help find out the answer.

The celebrated scientist Erwin Chargaff was an Austro-Hungarian biochemist who immigrated to the United States as he fled the Nazi era. He is best known for research at Columbia University in the mid-fifties that paved the way for the discovery of the genetic role of DNA and its method of replication. He left a well-loved memoir, *Heraclitean Fire – Sketches from a Life before Nature*.

In that book of reminiscences, Dr. Chargaff recalls being invited to Moscow in 1957 for a scientific conference. Before leaving for the Soviet Union, he was visited in New York by a strange woman he nicknamed "Mrs. Grizzly." He says she looked "very much like an underpaid mother with several difficult children."

He adds:

> "At this occasion, no money was offered: everything was strictly scientific, although the poor woman was clearly out of her depth as a spymaster. Mrs. Grizzly asked for help with difficult questions which she had trouble spelling correctly. It was not clear why she had come. She was pleasantly confused, exuded warmth, wished me a pleasant trip and added, unfortunately, 'See you again'..."

This was the summer of 1957, and the spooks' request for Dr. Chargaff to "keep his eyes open" had the opposite effect: Not eager to become some sort of spy, he studiously avoided contact with his soviet colleagues. He visited Russian museums and made sure he learned nothing.

Back in New York, he found Mrs. Grizzly waiting for him: "Her people wanted to know whether the Russians had succeeded in creating a homunculus, *and tiny men with whom to populate their spaceships.*"

Throughout this research, like Mrs. Grizzly's CIA bosses, we kept looking for indications of human trickery: Could the "avocado" have been part of a hoax by the Russians or even the Germans, as an ultimate sign of defiance? But the short beings couldn't be explained that way. Nor could the extraordinary correlation with events at Socorro and at Valensole. So we made renewed efforts to locate additional data to confirm the core story.

∧ ∧ ∧

CHAPTER EIGHTEEN

THE FIFTH WITNESS

The breakthrough came in April 2022, when persistent research by Paola Harris disclosed the details of the Padilla household in the years following the recovery of the 'avocado'. Those details allowed her to fill the blanks in the later story.

Sabrina was not the only child in the Padilla household in the 1960s: Paola found out that grandfather Faustino Padilla had adopted another child left stranded in the turmoil of the war and scattered families. That boy, Faustino Myers, was Sabrina's cousin, the son of Gertrude Castillo, Jose's sister. Mr. Padilla raised him as his own kid.

He had kept a good memory of the strange material around the house.

We already knew how Sabrina grew up in that post-war environment and witnessed the aftermath of the crash, both in terms of the landscape changes and in the metallic artifacts the family had recovered and saved.

The reader may recall that one of the most puzzling of these artifacts was the kind of fiber Jose Padilla and Reme Baca had called 'angel hair.'

There had been a lot of it scattered around the communication tower and elsewhere. As we saw, the kids had saved entire bags full of the stuff, ignored and left behind by the soldiers.

The fibers were of particular interest for Paola and me, because both of us had known Colonel Corso and separately interviewed him at length about the period when he was in charge of the UFO crash materials gathered by the Army Research services after World War Two, and distributed for study to trusted branches of American high-tech industry. We knew those secret materials included something the Colonel called 'fiber optics,' although he had conceded to me in Las Vegas that it didn't look at all like the 'normal' optical fibers one could now find in any communication network and in most modern American homes.

For one thing, the fibers described by Colonel Corso glowed in various colors along their whole length. For another, they did not resemble long wires but they presented in tight bundles. Those were the exact characteristics of the devices Jose, Reme and later Sabrina had picked up, wondered about, and used as toys or as ornaments for their Christmas trees.

For Paola, it was time to pick up the phone and call Sabrina again, to document what Faustino Myers might have seen, and to find out if he could support the body of testimony we had already gathered.

The conversation, early in April 2022, went like this:

Paola Harris: You were starting to say, Sabrina, that you asked your uncle Jose (Padilla) why there wasn't a Christmas tree, back then?
Sabrina Padilla: I asked him, was there a Christmas tree in the house, when he was there? He said yes, they had Christmas trees every year, him and his Mum and Dad. They decorated the tree every year. But when I came along in 1953, they didn't celebrate too much (with) Christmas trees there (any more). It's probably my cousin (Faustino Myers), who put one in his room, but we didn't have one.
Paola: OK, now when Jose remembers the Christmas tree, he used the 'angel hair' on it, right?
Sabrina: Yes.
Paola: So, (you told me) the angel hair was in a plastic bag. Please describe the drawer where it was, and what else was in the drawer.
Sabrina: Well, that was MY angel hair. The one I had in a plastic bag was my angel hair, personally.
Paola: Where did you get that from?
Sabrina: Oh, my grandmother (Iñez Lopez Padilla) gave it to me.
Paola: From the original angel hair, right?
Sabrina: Yes, 'cause I just grabbed it, and it was in a little piece of, uh, plastic. And the rest of it, I don't know what they ever did with it. In

Chapter Eighteen - The Fifth Witness

fact, I think, I might have been the last one that had it, that bag, and I put it away with my pigtails.

Paola: OK, how old were you when they cut your pigtails?

Sabrina: Probably about seven or eight, something like that. (Note: that would place it about 1961. Her cousin Faustino Myers was 18 by then).

Paola: And they cut your pigtails and they put it in a drawer, and where was the drawer?

Sabrina: You know, the way that those houses are made, over there, the drawer was in the room where you go in, through the front of the house (...) It was a drawer where we kept clothes and things like that, you know, money or purses, underwear...

Paola: Ok, and that was your angel hair, you considered it yours?

Sabrina: Yes, because I had it with my pigtails.

Paola: So, there was someone (else) living in the house, and that was (your cousin) Faustino Myers. But it doesn't mean that his Christmas tree had the angel hair on it.

Sabrina: It DID have the angel hair on it! But it didn't come from the bag that I had mine in. It was from some... from one of the *viroccas* (storage rooms) back there, where we had so much of it.

Listening to the tape, I grew impatient as the discussion went into such detail, but that level of precision was necessary so that memories drawn from the past could be re-awakened and impressions checked against later events. When Sabrina came along in 1953, her cousin Faustino Myers would have been ten-years-old. The incidents about the unusual fiber materials would have taken place seven or eight years later, when Sabrina's cousin was just starting his own family.

Paola: He was living (there) with his wife, and how old was he?

Sabrina: Oh, he was about ... they were really young. I think he's 79 now.

Paola: So, they were a young couple. Were they married?

Sabrina: Yes, eventually they had three children, as they went on, you know...

Paola: But did they have like, one baby when you were there?

Sabrina: They had two of 'em later on, as time went by....I don't really know how old they were then, I never thought about it (...)

Paola: OK, when you remember seeing the Christmas tree, how many children did he have?

Sabrina: Ok, lemme see, that time, might have been just two? And (later) they had a little girl named Lisa. She's buried right next to my mom over there at the cemetery. And, you know, she was the last baby. It was just the two boys at first.

Paola: So, there was a room, next to yours, with Faustino Myers and his first wife, and there was a Christmas tree in the room.

Sabrina: Yes ma'am. Yes ma'am.

Paola: And it was in their room, because that was kind of their part of the house (...) Did you ever go in their room?

Sabrina: Yeah, I went in the room, that's why I knew it had angel hair on there.

Paola: Oh, they let you come into the room, but that's also where they slept?

Sabrina: Yes, that's right. And they would close the door for privacy, 'cause the way those houses were built, there was no hallway. You would go from one bedroom to another, to another. It was like, nothing to divide it, other than a door between you and the bedrooms, you know.

Paola: So, you would go in there, they'd let you go in, at 7 or 8-years-old. They'd let you go in and see their Christmas tree, and it had that angel hair. Did they put it all over the tree?

Sabrina: *All over the tree, yes ma'am.*

Paola: Well, they must've got it from (your grandpa) Jose's stash of angel hair, they couldn't have got it from anywhere else.

Sabrina: They probably got it from my grandpa's, you know, in the back over there, in the rows they called *viroccas*. They had it back there in a gunny sack, and they probably just got it and put it on there for decoration. And *I remember how beautiful it was, oh, it used to just glimmer.* You didn't even think, you know... To me it was so beautiful, *that you didn't even think about the decorations, it was just so beautiful on the tree...*

Paola: Did they put it anywhere else in the room?

Sabrina: No, just on the tree. I remember that! His wife gave me a little baby doll that had a yellow little pyjamas on it. I was happy to get that present!

Research is all about people, I thought. Building trust and taking the time to understand what brought about the experiences, even something as

Chapter Eighteen - The Fifth Witness

simple as the sense of wonder at unusual light colors or reflections from an especially festive family gathering at Christmas time. Paola picked up on those recollections:

Paola: We didn't realize, Jacques and I, that there was anyone else in the house that could have witnessed this. So now, to confirm all this, we are looking for Faustino Myers, who is possibly 78-years-old. And his second wife, Esther, has been keeping in touch with you?

Sabrina: Oh, for many years!

Paola: And, when did you meet his second wife? Were you in good rapport with young Faustino?

Sabrina: Me and him have never been on bad terms. To make a long story short, his mother and my mother were sisters so they left me and him there, and my grandparents raised both of us.

Paola: That explains everything! What is his mother's name?

Sabrina: His mother's name was Gertrude Castillo.

Paola: His mother is Jose's sister, isn't she?

Sabrina: Absolutely.

Paola: So, two of Jose's sisters had these children. They had Faustino Myers, and they had you.

∧ ∧ ∧

Without the detail of such dialogues, a subject as unique and diverse at the UFO phenomenon doesn't make sense, just as what we so often see on TV or on the Internet—a few minutes of information extracted from breathless interviews about strange objects passing in the night--doesn't make sense.

Such media reports try to capitalize on the thrill of the unknown, but they add nothing to our comprehension of a truly important set of unexplained observations, whether the witness is a Navy pilot, a housewife or a farmer in the field looking for a cow: We couldn't go on with the above reconstruction as our only base, so Paola resumed her dialogue :

Paola: Now that we know Faustino Myers was in the house, it's very important for us to find him and to ask him, if he is still alive, about the angel hair.

Sabrina: Yes ma'am.

Paola: So, he doesn't know about the crash, or does he?

Sabrina: Apparently, when I talk to Jose, he doesn't seem to know about that. He really didn't care about that (...) Or it just didn't amaze him. I don't know what else to say. Maybe he just didn't care about it.

Paola: Well, he probably didn't know. Not everybody knew (...) Maybe they didn't talk about it as a UFO crash or anything, at that time.

However, because he was in the house, and because he touched the artifacts, do you think he also touched the aluminum (parts)?

Sabrina: I don't know. I don't think he ever, to be honest... I don't think he ever touched the 'aluminum'. He wasn't there when we did that. That was way back. He was gone already, or it was one of those times where he moved out of the house and they moved into a trailer, and later went away. So, it was just one of those things...

Paola: OK, so do you think that we may be able to have him talk to us about the angel hair?

Sabrina: Oh, I would love to. I haven't seen him in over fifty, sixty years, and there's no reason why he wouldn't want to see me, cause me and him didn't part with no hardships. We were close like brother and sister, and we were both left over there and my grandparents raised us both. He finally changed his name from Padilla to Myers. That was his father's name. His father was a white man.

∧ ∧ ∧

After that conversation with Sabrina in the spring of 2022, Paola had to do quite a bit of research to join Mr. Myers and his second wife Esther at their home in New Mexico: everybody had scattered and it was difficult to reconstruct what had happened to the various branches of the family until modern times, but arrangements were finally made to meet the Myers for breakfast in Socorro, and to spend several hours going over details of everybody's lives, way back in those difficult years when the country was still trying to recover from the war, the atom bomb devastation, and the tragedies so many families had endured.

We had a hard time connecting with Mr. Myers. Communications were difficult, the initial rendezvous in Socorro was missed, and it took multiple phone calls around the neighborhood for Sabrina and me to finally

Chapter Eighteen - The Fifth Witness

locate Mr. Myers and arrange to meet. Once we did, he very kindly invited us to his house, less than an hour away, in a small cluster of homes sheltered under trees in the beautiful New Mexico landscape.

Happy to see visitors, several excited dogs were barking and jumping around us as we made our way to the front door. We noticed lines of colorful rocks arrayed in several rows along the path, some of them striking in shape and texture, reminding us of the geological diversity of New Mexico. Evidently, Mr. Myers is a knowledgeable collector of mineral samples and petrified wood.

The conversation started with reminiscences about the time of the war and its aftermath. I told Sabrina: "When you talk to people these days, young people – like your grandsons, they have no idea of what things were like, back then..."

Sabrina: They don't know. They don't know how to cook without gas or a propane stove. They don't know how we had to chop wood and all that stuff. Even use the outhouse, OK? You know what an outhouse is?

Jacques: Yes, yes, I know!

We all laughed, and conversation turned to the sheer tranquility and the appeal of the surrounding landscape. I said: "It's wonderful here. It's so peaceful..."

Sabrina: So peaceful, it's beautiful, yeah! OK, (to Faustino) you said that house is for sale over here. They'd probably want a lot of money for that house. Yours is a beautiful house you have here! It's gorgeous! When can I move in? Where's my room?

Everyone laughed again. Sabrina looked around the room, commenting on how nice everything was and on the convenience of having three bedrooms. But what about the challenge of the high desert, with its extreme swings in temperature?

Faustino Myers (pointing to the wall next to him): This wall is a heating wall.
Sabrina: Oh, the heating wall! OK, I've seen those before.
Faustino Myers: Right now, we don't get any heat, because the sun is not

over us. But in the summertime the sun comes over there, shines on it, and then in the night, it releases the heat.
Sabrina: Yes, yes, the heat, like an adobe house.
Faustino Myers: Yes, it's adobe.
Sabrina: It's gorgeous, I love the wood and everything.
Faustino Myers: That other big one over there, it's got a heating floor. It radiates.
Jacques: Where do you get the water from?
Faustino Myers: Drinking water? We have a...(hesitating)
Jacques: From the company pipe?
Faustino Myers: Yeah, the pipe.
Jacques: You don't use a well?
Faustino Myers: Uh, we don't have one on this one. At that other place, where my daughter lives, there's a *chello well*, they call it.

In the days of the crash, few people had electricity and all the water would have come from wells. The reader may recall that, back in San Antonito, the Padilla home drew its water from a well, and there was a tall windmill tower next to the house to provide the flow.

Sabrina explained I had written a book about "those days," and the time when the first atomic bomb went off less than 30 miles away. I commented on how hard it was now to get people to understand what things were like, back then.

Faustino Myers: I didn't get over that, I guess, because I still chop wood and ... The only thing that I have, that I can't get used to, is living in an electric place. The stove is electric, where the other one was propane or wood. This one's too hard to learn how to cook.
Jacques: What you guys went through...
Faustino Myers: Some of my older members of the family...not many's left. And they lived at southern San Antonio, and the lady was related to my grandma and the man was named Forassio Padilla and they lived over there. He had, like, twelve kids, something like that.
 When all these things happened, it left them all bald. I was the only long-haired guy in the family! Everybody would get mad and try to cut my hair off. And they claimed, because of the bomb.
Sabrina: That's what happened with my fingers, see?

Chapter Eighteen - The Fifth Witness

Fig. 31: Faustino Myers and Sabrina Padilla as children, circa 1957.

Fig.32: Witness no. 5: Faustino Myers' testimony, April 2022.

Faustino Myers: And then the women and the guys, they were all sterile. Nobody could have families.

Sabrina: I was only able to have one myself. My whole life, just one child and then I lost her, so I don't have nobody. I go over there to see Lisa over there in the cemetery. Lisa was my baby.

Well, I wanted to tell you that Jacques is interested in the Christmas trees we used to put up over there. You remember the little Christmas trees we used to decorate with all that stuff?

As we were about to talk about the crash and its aftermath, leading to the recovery of various unknown materials including the fibers with their colored lights, Mrs. Esther Myers, Faustino's second wife, arrived at the house and joined us. We thought she could add precious insights to what Mr. Myers recalled. The introductions were festive:

Sabrina: I'm happy to see you! It's been too many years! It's been too many years!

Esther Myers: We just know each other by our signature! (laughter)

Sabrina: This is Jacques Vallée. We work together and he's doing a story on the Bomb, and what happened here in New Mexico...

Esther Myers: Oh, yes.

Sabrina: In White Sands, New Mexico.

Jacques: I'm writing a book about that time. That whole time, because people....

Esther Myers: Looks like they are making a little headway, those people – the *down-winders*. I saw it in the news somewhere, that legislature was considering some compensation. You know, for those people who survived.

Jacques: Well, it's about time!

Sabrina: But some of them might not live to see it.

Esther Myers: Some of them are already dead from their cancers, you know.

Sabrina: I've had four operations myself, beyond that madness.

Esther Myers: So, you guys did fly in?

Sabrina: Yeah, I flew in. All of us flew in.

Esther Myers: I see, you came in yesterday with all that wind? My God, how did you guys land?

Sabrina: It was the worst weather, yesterday. I don't know how we landed.

Chapter Eighteen - The Fifth Witness

Jacques (laughing): We still have some bushes stuck in the front of the car.

Esther Myers: I imagine! Yeah, it was horrible. It was horrible. Well, I'm glad to see you.

Jacques: We were talking about that Christmas…

Sabrina: The Christmas…you remember, you used to decorate the tree and things like that, you know?

Esther Myers: Yeah.

Sabrina: Do you remember the name, the little *pelo* that we used to put around it, you know, we used to call it angel hair?

Esther Myers: Yes, the angel hair. Yeah, I remember.

Sabrina: We're doing a little study on that, 'cause we found it to be strange.

Esther Myers: And then it hasn't been around for a long time.

Sabrina: No, it's not around. I think my grandfather had it in the *virocca* somewhere back there, you know, why did he have it in the *virocca*?

Faustino Myers: Right.

Jacques: People don't remember that time, and there is nothing like it now.

Sabrina: I wanted to ask this question, do we have your permission, and yours, to write about that, about that you seen the angel hair of the Christmas tree? Do we have your permission?

Faustino Myers and Esther: Sure, yeah, oh yeah.

Sabrina: OK, 'cause we just want you to know that, when we can give you the book when it's written, you know…

Esther Myers: It takes a while to get it together, huh?

Jacques: We've got most of it already together. I just want to get as many…, as much report on it, to try to understand it all. I live in San Francisco now, so I talk to technical people, scientists and so on, and they can't figure out what that material was. Because you didn't have any (electrical) power, right? *You didn't plug it into anything?*

Faustino Myers: No.

Sabrina: It kinda shined, remember…

Jacques: What else - can you describe what it looked like?

Faustino Myers: *It kind of glowed.* Yeah, yeah, it was, it was…

Sabrina: See, it would bite you when you touch it.

Faustino Myers: *It was something on one's hand, it was just a frenzy.*

Jacques: Did you see that in the dark, when you were completely in the dark, did it keep glowing? What do you remember?

Faustino Myers: I just remember that when you touched it, it was something like fiber glass, *it would poke you*...it was, uh, you know how not to get fiber glass because it gets in your hand, you know....it would do that.

It occurred to me that Mr. Myers' indifference to the crash of the avocado, perhaps due to the fact that Jose and Reme, the two boys who had witnessed it, were gone, and that Jose's father was consistent in his attitude of secrecy, that indifference itself was actually an asset. Mr. Myers did not relate the strange fibers to a peculiar event at all. And this made him a more important, unbiased witness to the handling of the material.

He had just mentioned that it looked and served as angel hair. What else could we learn?

Jacques: What was it like? Like what we put now on Christmas trees, with those lights... or was it a bunch of (other) things?
Faustino Myers: It was like an icicle.
Jacques: With a bunch of fibers?
Sabrina and Faustino Myers: Yeah, yeah.
Jacques: It wasn't a thread? Or was it like a thread? (to Sabrina): What you remember is something that was like a clump that you didn't have to plug into anything.
Faustino Myers: Uh, huh. Yeah, we just remember, uh, *well they poke you...*
Sabrina: Yeah, it sure did. It would shock you, but even though it shock you, it still feel like it cut you.
Faustino Myers: Yeah, it felt like that.
Sabrina: You wouldn't want to touch it too much.
Faustino Myers: It felt like that.
Jacques: Where did you get it from?
Faustino Myers: Um, it was in a shed at our house, I guess, *I don't know where it came...*
Esther Myers: Where did you think your grandpa collected it from?
Faustino Myers: I don't know. I would go and take some up when nobody was looking.
Jacques: Was it glowing by itself?
Faustino Myers: Yeah, yeah.

Chapter Eighteen - The Fifth Witness

Sabrina: Shimmering look?

Faustino Myers: *It was different: little colors there, but to me it looked like it was hair.* For me it's like that, uh, what do you call that, what you put in the (Christmas) trees…?

Sabrina: Icicles?

Faustino Myers: Icicles, it was like that, *but it was real thin, not an icicle thin.*

To me, the dialogue was made even more remarkable because Mr. Myers had not been involved in the 'avocado' incident, which was still a family secret for the older members. Jose had left the home as a teenager to join the Army, and again, we already know that his father kept his own secrets about the recovery of the craft. In that first conversation, Sabrina and I didn't want to raise the matter of an 'alien' origin for the glowing fibers, in order to accurately record what Mr. Myers remembered.

Jacques: So, the other people in the house, where did they say it came from?

Faustino Myers: Well, I don't know, *we never talked about it.* It was in a box and then we decorated our trailer, but I don't know where they got it, I don't know.

Jacques: What happened to it?

Faustino Myers: Well, when the house burned down, everything burned there. But that stuff, it burned, and it knocked the color off, but it stayed like in a string, you know?

Jacques: Uh, huh?

Faustino Myers: *It didn't burn all,* but it just, they didn't… *it didn't dissolve…*

Jacques: It didn't melt?

Faustino Myers: No.

Esther Myers: So, that fire, when your house burned down, it was in the shed, that stuff, and the shed burned down too?

Faustino Myers: Yeah, yeah. Boy, I'd have to think back a long…

Jacques: Was it the same color, after the fire?

Faustino Myers: It got blacked, you know, but it was still there.

Esther Myers: They must have collected it from somewhere, your grandpa and his *compadres* or whatever.

Sabrina: You know, my grandfather was somebody else, you never knew what he collected!

Now it seemed everybody had gone back to that time, after the war and the bomb, when strange things were still happening and local families were trying to recover economically and emotionally from the trauma of the fighting, and all the deaths. For families within range of Ground Zero, there was also the lingering effects of the atomic bomb.

Faustino Myers: My grandfather, his brother, one, two, three, three brothers, one sister – five of them, and *they all died of cancer*. It hit the whole family. And then when children came up, Marcelino's children, the first youngest one of the children died from cancer– she was 18-years-old. And then cancer killed, or I said, cancer put Diana away– and then her mother had cancer.
Sabrina: Yeah, I remember Diana.
Faustino Myers: And then Ruby had cancer, and then everybody in the family had, except me, I guess. I don't have any, I don't think.
Sabrina: We're the strange ones! We're still hanging out.
Esther Myers, to Sabrina: You've had symptoms?
Sabrina: I had four operations, cancer operations.
Esther Myers: Oh, gosh! And you were about two-years-old when that bomb set off, huh?
Faustino Myers: You're right, yeah.
Esther Myers: But you always lived there, until….
Sabrina: Do you remember the year of that Christmas that you put the Christmas tree up and you used the 'hair' on there? Do you remember the year? It had to been in the '60s maybe, huh?
Faustino Myers: Or lower than that, it's gotta be in the '50s: '57, someplace in there. Well, I was about two-years-old when the bomb exploded.
Sabrina: So, you remember that, huh? I came eight years later.
Faustino Myers: But I think it was March, when it blew up, because they just had that… (The actual date, as we know, was July 16, 1945).
Esther Myers: That anniversary, yeah, they did. Open House at Trinity Site.
Jacques: I've been there. It's impressive.
Esther Myers: Have you? It is impressive, when you think about it. Yes. It's bare, but gosh, you think, what happened there…

Chapter Eighteen - The Fifth Witness

Faustino Myers: When all that thing happened when we were small, but we had animals, farmers had animals there and all that, and all the animals were, they had, like the flu.
Esther Myers: They got sickly, huh?
Faustino Myers: The chickens were... their noses were running, their eyes were all gluey, and they had all kinds of saliva...

The animal control officers had told us the same thing, and the ranchers remembered it as well. But nobody from the Defense department, or from other branches of the government, bothered to tell the truth to the population. Farmers and ranchers were left to fight the diseases by themselves; including the illnesses in their own families, and losses among the cattle.

Esther Myers: Lot of mucus, their bodies were getting rid of a lot of mucus.
Faustino Myers: And we were, from where we were at, to that thing, we were about 45 miles away. And then, uh, some of it, well it went everywhere. But the B* family over there, and Fort Craig, they were closer, it was an open space to where they went. A lot of them, they were sick with cancer. And one of my aunts that got sick like that, her and her husband, trying to have children and all that, and she became pregnant but when she went to the doctor to check out what she had, the doctor (said) she had, like grapefruits...
Esther Myers: Tumors, there was no baby, it was tumors.
Faustino Myers: ..., inside, like she was a pregnant lady and the doctor took them and told her it could have come from that. And then the doctor had told her, "you can do whatever you want to, but you're going to have to fight for a long time for the Government, before ..."
There's nobody left but guys like me in this age, Cliff from there, everybody else that had it, they're all gone, you know.
Jacques: The explosion was four times more powerful than they had calculated.
Esther Myers: Oh my gosh, really? I didn't know that.
Jacques: I found the book by the wife of Enrico Fermi. She worked at the hospital in Los Alamos and she knew some of the things that were going on, even though the bomb was secret. By today's standards, it was a small bomb, a small thing. The generals didn't want to use

it, but the war could have kept going for another three to five years. So that's what I want to write about, and I really thank you for what you remember. It's very valuable because people have no idea; and those fibers, we still don't understand what they are, where they came from.

Faustino Myers: They came after that. Us, we growed up and all that. When they were gonna blow something up (they haven't done that in a long time) but when they were gonna blow up anything, they would make everybody leave from over there. Get out of there!

Sabrina: But the most unfortunate thing was they didn't give the option, when they dropped that bomb (note: in 1945).

Esther Myers: Right, it was a surprise. Nobody had any cover. And those particles, I guess, they go everywhere.

Faustino Myers: And then they throw out, they go, we could feel that, we could hear it. BOOM! What one rancher told us, where they lived over there… They lived, much closer to that than we did, and he said when they'd done all of that, in that area, the ground was white, like dust. They had a *burro* that was black and he turned white. It made him white. It made him white, and all kinds of stuff like that.

Esther Myers: Poor animals!

Sabrina: They suffered, you know. So, imagine what that did to our bodies, for those who lived over there too, you know.

Esther Myers: Well, I know that you folks have some relatives of *R*, and his wife…They had about six children and five of them could not have kids. They tried, only one of the girls was able to have kids. Everybody else tried and tried and finally they ended up adopting. And I think it's because of the bomb, that's what I think.

Faustino Myers: But most of that stuff now… I worked there for a while, I was hauling petroleum and I helped fuel, and all that, and they were still doing a lot of stuff.

There was another knock on the door and a young man, Mr. Myers son-in-law, came in with his wife and two laughing kids. It was time to enjoy the company, and leave the conversation about the 'avocado' and its strange properties for another day. But I couldn't forget what I had heard: the continuing sad, colorful and puzzling story of those events of August

1945, which nobody in government or in academia wanted to acknowledge, those peculiar physical effects no one seemed ready to discuss...

One thought occurred to me, as my plane lifted from Albuquerque on the way back to Silicon Valley, recalling the immense sadness in the voices of the people we had just interviewed about the aftermath of the bomb: *This isn't about pretty lights in Christmas trees.* It's about truth, and life itself, and the responsibilities of our elected officials.

If the people in place in the highest levels of government never told the citizens of New Mexico about the devastating effects of the nuclear explosions back in the late 1940s, why would they come clean, any time soon, about the UFO phenomenon and its meaning for humanity?

^ ^ ^

CONCLUSION

The universe contains no anomalies, and the appearance of an anomaly is warning that our understanding is inadequate.

Mathematics must be the servant of understanding and not its master. Equations, by their very nature, cannot discover, they only yield relationships derived from the initial statements which were inherent in them when they were stated.

Truth is not hidden, it is available to all and is the same for all. Apparent differences must be due to inadequate understanding.

Canadian Government UFO expert Wilbert Smith, 1961

CONCLUSION

Perhaps a large part of the activity that's classified as UFO activity—abduction and the whole host of this type—may very well not be due to extraterrestrial activity at all.

I would suspect that if any is due to ET activity, it is a rather small part. And a larger portion is due to human type, Earthling type activity in a very clandestine fashion.

Captain Edgar Mitchell, Apollo 14 astronaut
Sixth person on the Moon

The full story of the crash on the Padilla Ranch is likely to remain a closely-held secret for a while. Paola and I are now confident of it, and as we complete this investigation, we know why.

So does Mr. Jose Padilla, who was there.

The fact that we have recovered and published his testimony, along with that of Mr. Remigio Baca, completed by those of Mrs. Sabrina Padilla and Mr. Faustino Myers, is the result of a series of chance occurrences that involve a simple, independent genealogy search. We hold a material artifact Mr. Padilla extracted from inside the craft, simple to explain yet loaded with long-term unanswered questions, but that is irrelevant to eventual revelation of the reality of the event.

That reality can continue to be denied or safely distorted for a long time.

The people who hope for an imminent "Disclosure" about UFOs, and who are making valiant efforts to document the phenomenon, should take one important fact into account: *Disclosure could only come from the same organizations that are in charge of the security brief itself.* And those organizations have their own agenda and constraints, which may be regrettable but possibly legitimate.

A government entity or its classified contractors can claim to know the answers and confuse the issues with impunity. It can even come up with fake disclosures to feed the public's thirst for what passes for "truth" and the media's excitement to promote it, in the interest of what may be perceived as a higher good.

We do have the right to ask: A higher good, for whom?

From a physical point of view, only a very small group seems to have an approximation of the real thing: Not the usual military or intelligence agencies so often accused of hiding "the truth," even if each of them does guard its own best information and hides some bits of material in the basement, like Jose and Reme did with their precious *Tesoro*. There is, literally, nothing to disclose *on that level.* Within the structure of Government, the usual security clearances don't seem to apply here, so Congress cannot send out the Federal Marshalls to force some scientist to come before a special committee and testify, or waive a former oath of secrecy:

What secrecy? Is anything written down somewhere?

When one puts it that way, it's actually funny. Or, as "Sir" Arthur Lundahl, legendary founding director of the CIA's Photographic Interpretation Center (APIC) told me at his home on September 14, 1989, when I was following up on an especially sensitive UFO case where the public, the media and the Academy had been told a huge lie (or rather, a series of untruths, still un-revealed) it is *ironic.*

So ironic even the most sanguine UFO researchers and podcasters refuse to touch it.

Any Disclosure will be managed by the same groups that have safely managed the dossier for the last 75 years, and they can spin it to their liking, claiming that the military must retain control. They can also re-position it in any shape and context they want through the multiple "cut-outs" they can create out of thin air to send us into blind alleys. That's what they do.

Nobody can put pressure on such an organization. If you're not cleared by them you can't contradict them. If you are, you're certainly not about to contradict them.

But the logical maze is a bit more subtle: *There may not be anything to contradict or to affirm*, in the sense of the way good people have been asking the question. At most, some bits of incremental technology may come out in time, through the normal intellectual property legal channels of patents and trade secrets, with the business awarded to top-tier, high-technology companies serving the Defense department and the space program, as Colonel Corso claimed.

We all know who they are.

Yet, even those companies may refrain from implementing such an arduous business plan. They may choose to put their smartest researchers on other practical, financially lucrative projects: We still use those big inefficient rockets weighing hundreds of thousands of pounds to send a couple of astronauts to the moon, carrying a small shovel and two days of dehydrated rations, don't we? As we did back in 1969.

There are a few clever physics experiments under way. Some friends of mine among venture capitalists claim to know where those labs are. But even our cherished Silicon Valley is a long way from practical-scale, reliable antigravity products; and even farther away from the well-documented performances recorded by professional observers like pilots or technical personnel during UFO events; or witnessed by little kids in New Mexico, playing among the cow pies.

∧ ∧ ∧

What the Padilla Ranch teaches us is that the real secret isn't readily operational, waiting for us in the vaults of the Intelligence community or the Pentagon, just waiting for some bureaucrat to pull out the *Declassified* stamp and triumphantly slam it on the document to the applause of an appreciating public; and even less at the White House, where political plots are hatched, and little more. If there is evidence for such a secret, it may reside with the long-term Keepers of the atomic knowledge, because they are the ones who had custody of the 1945 craft seen by Jose and Reme.

They would *appropriately* follow their own different, parallel system of clearances at the highest level: the nuclear clearances, as opposed to the ordinary, national security clearances that thousands of government employees and industry technicians hold. It would be their baby, after all; delivered to their doorstep, less than an hour from Stallion Gate, just one month after "their" bomb.

Nobody would want to acknowledge that. Or need to.

We shouldn't even fantasize about a cabal, or a secret government, or a Vatican plot, or a grotto of Bavarian Illuminati dressed in funny robes. More prosaically, the Keepers of the Secret (I prefer to just call them "the Custodians," because they may not have gained any real understanding of what they are supposedly guarding) may be a specialized team that cuts across the stovepipes of classified technology development and needs no points of regular interaction with ongoing Intelligence projects, pontificating Academics, or Congressional envoys, *because they are self-contained.*

The only thing elected officials would have to know about them is the fact that they need to be left alone, beyond the purview and oversight of most legislative and judicial bodies.

And that has been the state of affairs for 75 years.

The situation is exemplified by what Dr. Louis Hemplemann, Laura Fermi's boss at Los Alamos, said in a 1945 statement (not revealed until 1986, however) regarding the effects of lingering radiation on grazing livestock and the ranchers who tend to the animals: He acknowledged that *"a few people were probably overexposed, but they couldn't prove it and we couldn't prove it. So we just assumed we got away with it."* (**44**)

To his credit, Dr. Hemplemann did officially request that the War department investigate the health of residents of Bingham, the closest town North of Trinity. But according to an article in *Health Physics* published in 2010, it seems the request only led to non-specific questioning under some pretext. "They were never told, *You might want to leave. You might not want to sell that cattle,*" observed Reme Baca. It was only in 1980 that the ranchers found out that the real reason for those visits by Army Intelligence had to do with observing *them*, to document the human effects of the bomb on them and on their kids.

Yet, much of the need for secrecy was over once the bomb had detonated. How hard would it have been to go around and tell the ranchers not to drink from their water tanks?

∧ ∧ ∧

Conclusion

Are there some facts that support that interpretation? Quite a few, as it turns out. We can look at five actual points of reference, as practical examples.

-A-

When high-level Canadian Government scientist Wilbert Smith, who arguably did have the *need-to-know* about the situation (his services had gathered their own hard evidence from crashes in Canada), was told that the real secret was held "higher than the Manhattan Project," implying it would be politically awkward to ask any more questions on behalf of Ottawa, he was respectfully but firmly presented with a simple truth.

Responding to an inquiry in 1958 regarding his own Project *Magnet* (**45**), Wilbert Smith did lift a little corner of the secret: "Various items of 'hardware' are known to exist, but they are usually promptly clapped into security and therefore are not available to the general public. Substances such as 'angel hair' and molten tin, etc. have been observed to drop from these craft, and have been gathered and analyzed. Strong magnetic disturbances have been observed in the vicinity of these craft. In fact, I would say that many more people have more evidence supporting the reality of the flying saucers than for the reality of atom bombs. But atom bombs bear the stamp of official disclosure, and in posting your representative with questions, please be sure the questions you ask are exactly what you want to know. If you ask, "Does the Air force have any saucer hardware?" you will get 'No,' truthfully. The hardware is not held by the Air Force." (**46**)

When he was asked "You say that you had to return (a piece of hardware), did you return it to the Air Force, Mr. Smith?"

He answered: "Not the Air Force. Much higher than that."

The reporter went on: "The Central Intelligence Agency?"

Smith chuckled: "I'm sorry, gentlemen, but I don't care to go beyond that point. I can say that it went into the hands of a highly classified group. You will have to solve that problem - their identity - for yourselves."

-B-

There was much more to the Canadian study and its implications for American Intelligence, but somehow it has been forgotten in the fallacious cloud of "instant news" from online podcasts and breathless TV specials the public now enjoys.

Speaking before the "Illuminating Engineering Society" conference in Ottawa on January 11, 1959, Wilbert Smith flatly stated that he had, in fact, had custody of a fragment ripped out of an unidentified flying object by a US Navy fighter over Washington, DC.

Researchers C.W. Fitch, from Cleveland, and George Popovitch, prompted a follow-on conversation with Smith, where he stated on the record that he had shown Admiral Wilson a piece from a small flying saucer that fell near Washington (DC) in 1952. The metallic fragment was twice the size of a man's thumb.

The USAF had entrusted him with it for a short time. This was not the only fragment he had examined, he added. He found it much harder than "our" metals.

The pilot had been chasing a fluorescent disk. A piece detached itself and the pilot watched it fall until it hit the ground. *It was retrieved an hour later*. It weighted about 250 grams. It was cut with a diamond saw for laboratory study. The fragment Wilbert Smith examined was about one third of the piece. **(47)**

Chemical analysis found some iron oxide along with an agglomeration of magnesium silicate in a matrix composed of particles measuring 15 microns.

As we saw above, when asked if he returned it to the US Air Force, Smith said no, he returned it to "a much higher level."

-C-

When I met with "Sir" Arthur Lundahl at his home on September 14, 1989, following my testimony at the Al Gore Congressional Hearings on crisis management, he fully confirmed the statement by Wilbert Smith that I have quoted earlier and told me what had happened on the US side **(48)**. The incident had taken place in July 1952, at the height of the famous "wave" of

UFOs buzzing the Nation's capital--in the process, violating the restricted airspace of the White House and Capitol.

At the time, the media were told all the cases were just visual sightings, and Dr. Donald Menzel, of Harvard College Observatory, a leading astrophysicist (and former mentor of Dr. Hynek) found it convenient to "explain" the cases before the national media as optical illusions, essentially mirages due to a temperature inversion.

Most newsmen in the US folded their investigations and accepted the statement that there had been nothing more than questionable radar confirmation. That belief still stands, even in the most recent documentaries about the phenomenon.

Yet, according to *both* Wilbert Smith and Arthur Lundahl, a piece of metal did detach itself and was studied.

And the real reason it "detached itself" was that the Navy pilot in question shot the Hell out of it after receiving the authorization to fire at the flying disk when it penetrated the most heavily protected area in the United States, a fact that has been kept from the highly malleable American media for well over a half century. **(46)**

For further confirmation of the existence of authentic samples, and of the analytical results, I cross-checked what "Sir Arthur" had told me through private conversations with a third authority, Dr. Marcel Vogel, celebrated in technical circles as the inventor of the magnetic coating of disk drives at the IBM Advanced Research Labs in Silicon Valley, an unimpugnable source like the previous two. He explained to me why it greatly puzzled IBM when the government confidentially asked the company to have the piece examined in his lab. Naturally, nothing was ever published.

- D-

I was in Dr. J. Allen Hynek's office at Northwestern University in April 1965 when he arrived back from Washington, where he'd confronted then-Congressman Donald Rumsfeld about his need for higher-level information about what the government actually knew. Hynek was distraught and angry, because he knew the subject was about to escalate in the halls of Congress.

"Doctor, you do NOT have the need-to-know," the future Defense Secretary had told him, showing him the door of his office--and throwing him to the skeptical wolves in the academic community and the hecklers among the media.

Donald Rumsfeld was a highly-visible US Representative from Illinois, a Chicago native, and a leading, respected co-sponsor of the *Freedom of Information Act*, all reasons that made the rejection especially puzzling and painful to Allen.

-E-

According to various reports, on December 16, 2002, the Director for Intelligence for The Joint Chiefs of Staff, ("J2"), Admiral Thomas Ray Wilson agreed to meet with one of my colleagues from the NIDS (**49**) research group, physicist Eric Davis (**50**). They spoke in confidence for two hours, in a car parked behind the EG&G building at McCarran airport in Las Vegas. There, Wilson reportedly explained to Dr. Davis how, at a meeting with Navy Lt-Commander Willard Miller and others on April 9, 1997, he had learned that there were indeed "deep black" organizations in the US actively studying Alien technologies and bodies.

Both Miller and astronaut Edgar Mitchell argued that those rogue Defense companies needed to be brought under the control of the government. That same question had been raised in the 1997 book *The Day after Roswell*, by retired Colonel Philip Corso, the first man who had been in charge of distributing technology of presumably "Alien origin" to American technical companies, details of which he had privately told to Paola Harris and to me, in separate private meetings.

The reports (all or parts of which are denied by the Admiral) go on to state that he suspected high-level fraud, or at least improper processing of classified data. Launching his own investigation, Admiral Wilson is said to have privately sought the advice of General H. Marshall Ward and former Defense Secretary Bill Perry, who suggested that he "look through the records of OUSDAT," the Office of the Under-secretary of Defense for Acquisition and Technology.

He did, according to Dr. Davis' notes, and this precipitated him into the subterranean complexities of a very unusual, "deep black" program that was set apart even from Special Access Programs that are, by definition, beyond the casual oversight of Congress and the armed services, except through very specific and highly regulated procedures.

Conclusion

The Admiral was forced to confront the discovery that a top Defense/Intelligence contractor just "happened to have" a very large budget to run a program that appeared to be setup beyond the law and in apparent violation of the Constitution. And it had to do with UFOs. Oops.

According to the record of that conversation, Admiral Wilson made three calls to the company: To the general counsel, to the security director, and to the program manager. Instead of begging for forgiveness, as might have been expected, all three became upset at the effrontery of such a call from the government, and even "agitated and surprised," according to what Dr. Davis heard.

It seems this was followed by a secret face-to-face meeting inside a SCIF, a heavily-shielded and protected suite of offices, where the three executives had to confess to the Admiral *they had between 400 to 800 personnel on the payroll at various times, involved in reverse-engineering of a technology "not made by human hands,"* and that progress was frustratingly slow.

When Wilson said he would file a complaint because they were running a program outside all legal structures of the American government's operational and budgetary oversight, they coldly told him to go ahead: "Do what you feel you need to do." Admiral Wilson did try, and he soon heard from a very powerful Congressional structure called the Special Access Program Oversight Committee (SAPOC).

The answer was not what he might have expected.

According to Dr. Davis, Wilson was told that if he pursued his investigation, three things would happen to him: He would never be named Director of DIA; he would be forcibly retired early; and he would lose one or two stars.

The Admiral understood the message and went on to serve as director of DIA from 1999 to 2002. And nobody has heard about it since then. (ref.: Richard Dolan, **51**)

Here again, while the situation seems absurdly beyond everything we have been taught about the workings of our government, the fact is that we don't know the nature of what is being hidden. The simple explanation for what happened to Admiral Wilson is that the project(s) may be handled as "reverse engineering," outside the scope of an Intelligence agency like the one Wilson was directing. In which case the rejection of his request was legal.

At this point we should continue to ask questions in the name of Science and Democracy, but we should not presume anything about the answer. If the secret is, as Wilbert Smith stated, "above the Manhattan project," there may be valid reasons for a procedure of exception, and it would all be within the law. Personally, I am reconciled with the idea of never knowing what those companies are hiding. But there is a price to pay for keeping secrets from the people, just as there is a danger for revealing them.

On the other hand, perhaps the secret is that those highly-classified teams haven't found out anything useful from the hardware they've captured and secured. As a disabused scientist says in *Roadside Picnic*:

"A monkey pushes a red button and gets a banana, pushes a white button and gets an orange, but it doesn't know how to get bananas and oranges without the buttons. And it doesn't understand what relationship the buttons have to the fruit. Take the So-So's, for example. We've learned how to use them. We've even learned the circumstances under which they multiply (...) But we still haven't been able to make a single So-So. We don't know how they work, and judging by present evidence, it will be a long time before we will."

This makes me wonder how much the Soviet Union had learned over the years, from its own research on UFOs, but more importantly it made me reassess our own approach to the matters at hand, namely the curious events at San Antonito.

∧ ∧ ∧

What nine-year-old Jose Padilla, a bright and gutsy country kid, experienced for a full week on his family's land in New Mexico, tells a much simpler story: He may not have had the need-to-know about national secrets, *but he happened to be there to watch the object crash*, which was even better; he had a second, very bright witness with him. And there was little the Army, the J2 or even SAPOC or the Pope could do about it.

Or even find out.

Our two brave young witnesses were the right men for the job: they knew how to hide and they weren't about to give up a chance to observe what was going on as a dumbfounded detachment of young recruits in Army fatigues did their best to recover little pieces of a contraption their own superior officers had never seen--or imagined.

Conclusion

Jose and Reme were witnesses to an unexpected dialogue of sorts, an eerie exchange of symbols between the brightest scientists in the world and something else, undoubtedly the product of another mind. That's pretty much all we can say.

The history of this research teaches us that most of the symbols in that exchange remain to be deciphered; the technical folks have remained focused on technical things, arguing about such current concepts as quantum mechanics or relativity, thereby missing the main point entirely.

Officer Ted Jordan, the former New Mexico State Trooper who gathered trace information so thoroughly in Socorro, put it best when an interviewer asked him, "What do you think it was?" He flatly answered he had no idea, adding:

"I've thought about it often over the last thirty years. (...) *I don't think anybody else knows what it is*. I've seen people that—if they can't figure something out and give you a truthful answer—they will give you a snow job; and I think that's what it was; I think the agencies involved, they didn't know. They're supposed to know everything.(...) And they want to protect the people from their own selves, I guess, by, ah, not divulging any information. But at the same time, they don't have any information; they don't know."

And "snow jobs" they did give us: from the "experimental weather balloon" to the mysterious "Air Force Pogo Stick" (my all-time favorite!) to the even more imaginative yarn that was put in the ear of author Annie Jacobsen, who mentioned it in her book *Area 51.* (**52**).

Over lunch in San Francisco, Ms. Jacobsen told me a story that was brought to her publisher by Defense Department people who stated it was all true, although supposedly it had been kept a very deep secret. The yarn went like this: "In July of 1947, Army Intelligence spearheaded the efforts to retrieve the remains of the flying disk that crashed at Roswell (...) and they found bodies alongside the crashed craft. These were not Aliens. Nor were they consenting airmen. They were human guinea pigs."

The tale goes on to "reveal" that the poor deformed children found by the Army were kept at Wright Field until 1951, and then sent to "an elite group of five EG&G Engineers" (**53**) working at the Nevada Test Site. And what was the big truth behind the deformed aviators, child-sized with strikingly big heads and other deformities? They were kidnapped kids from Auschwitz, who fell in the hands of the evil Nazi doctor Mengele (The fact that he was an evil Nazi doctor is not in question here!... JV) who were subsequently re-engineered by him to turn them into grotesque child-size aviators for Stalin, in a secret deal that would allow Mengele to seek refuge in the USSR as the full defeat of Hitler became obvious. The kids were supplied to Stalin, who (being an evil genius himself) then reneged on the deal to save Mengele but kept the fake Aliens to play a trick on the US and dump them on New Mexico to create a panic in America similar to what followed the radio broadcast of *The War of the Worlds*.

Never mind that there was no actual panic after that broadcast, the supposed "panic" being manufactured by the media as part of the show's promotion, and the sociologists swallowing the hoax, and re-telling it ever since, because it sounded "scientific."

If you, gentle reader, have patiently followed this twisted yarn, you may have noticed that the Mengele story makes no sense historically or biologically, and it doesn't even make any sense in terms of the record of UFO sightings in the United States. Mengele was running for his life at the time, hunted by Israel and every Intelligence agency in the world, and he had very little time, and no facility, to conduct biologically absurd experiments. While that is supposed to explain the Roswell event of 1947, it fails to account for everything we know about Roswell, and of course it doesn't hold any value for the much earlier San Antonito crash of 1945. If you believe the story, we have a giant sardine from the harbor in Marseille that we can sell you.

Perhaps Stalin and Mengele didn't know about Jose Padilla and Reme Baca? They certainly didn't know about Area 51: The base didn't even exist in 1945.

The only real mystery is why somebody, in today's complicated technical world, finds it politically astute to go on inventing such "explanations," and planting them in publications sold as factual reporting to the unsuspecting American public.

∧ ∧ ∧

After a long time pondering the phenomenon in the field, in libraries and in the lab, Paola and I can only concur with Officer Jordan who "doesn't think anybody else knows what it is." But the phenomenon has a multi-level information structure that could be exploited. It has never been taken into account by scientists.

Conclusion

Let me try to use an analogy to explain how I believe that structure works, because the deception (or the ultimate key) is hidden within the illusion of discovery.

When you were a cute toddler, your mother, as the ultimate authority in your world, may have disclosed a big, urgent secret to you: She said that a small non-human being, a magical character, a clever Rabbit, had come out of his secret lair below the ground; he had gone around before dawn with a basket of bright sugary treats and had scattered them across the backyard.

This was a rare event, she added: He did that once a year, on Easter morning, but only if children like you had been good.

Now he was gone, of course, so you couldn't see him. You *might* have seen him, mind you, if you had taken the trouble to get up early, but you didn't, you lazy bum, so you missed the opportunity to ask him about any details, and anyway that would be demanding too much of such an important personage. But there can be no question he really does exist: *We have proof!*

Indeed, your trustworthy mother goes on, he has left "evidence." If you look cautiously through the bushes in the yard, you might discover some chocolate eggs and many candy delicacies wrapped in bright silver paper. Since this is a very special occasion, you are allowed to eat as many as you can find. And the best part: It hardly even cost anything.

The future will be wonderful.

∧ ∧ ∧

We have told this story because we believe the public's expectation of "disclosure" is a fallacy, just as our childhood hopes of seeing the Magic Rabbit was a (happy, warm, reassuring) fallacy. In the last few years, the Keepers of the Secrets have created a few legal and social instruments, media tools through which they can control leaks of information when they occur: The most practical channel to disguise and redirect such a leak is to encourage public "research," outwardly dedicated to genuine investigation of UFO reality, yet sufficiently embedded in the secretive structure of the modern government bureaucracy to continue shielding the most threatening or egregious aspects of whatever may be accidentally uncovered.

Recent history has shown that misleading revelations and even blatant failures of investigative practice in the pages of the newspapers or in the releases of the Academy of Sciences are generally tolerated by the public in the absence of courageous rebuttal, at least in the United States where social credulity before revered institutions has remained fairly strong.

In the case of the UFO phenomenon, this outwardly benign misinformation practice, based on re-direction of facts one can no longer hide from the public, may be well-intended. However experience in prior domains (the outrage of MK-ULTRA, the public health disasters of nuclear contamination, unsafe chemical plants and military disposal areas, or even the cover-ups of medical errors and pharmacological abuses by industry and government) has shown that delays in public awareness only contribute to distrust of government and to deteriorating image for major companies, even as it minimizes the immediate reckoning and painful legal consequences for the abusers.

Such misinformation is a pollutant in the civic discourse and the pursuit of human research, and an affront to the public's legitimate right to share in the knowledge of government officials, when it can have an impact on the conduct of life. The resulting distrust puts all of us in danger.

In the case of the crash at Trinity, it is science itself that is being abused: Basic facts are hidden from academic scrutiny and the actual reality of ongoing studies buried in black budgets continues to be denied, even in the face of official inquiries, as Admiral Wilson found out.

At the end of this series of investigations, we can only leave the research open-ended. We don't know what happened to the three creatures seen by Jose Padilla; we don't understand what an ordinary low-tech human device like the stolen bracket was doing, attached to that inside panel; we have no data to elaborate hypotheses about the propulsion of the craft.

In fact, scientists reading this book would be justified to ask, "What kind of Aliens would come from the other end of the galaxy, only to bump into a radio relay and crash in the shrubbery?"

To which we might add, "What kind of interplanetary spacecraft would contain a device made of a common, human industrial alloy, without brand identification, manufactured to precise metric dimensions (30cm long by 9 centimeters high) and metric diameters for all the holes?

But perhaps the device had just been installed by the Army grunts who cleaned up the craft? Possibly, but where did they find it? The US military wasn't using the metric system in 1945.

There is room for someone to suggest that the craft was a human device: perhaps the insectoid creatures were a weird product of the extreme

Conclusion

medical experiments carried out by several countries (including the US, by the way) in the insane days of World War Two, as Annie Jacobsen had suggested at Roswell. But what about similar "impossible" craft and "impossible" beings reported at Valensole and Socorro, and in other lands?

The facile explanation ufologists continue to put forward—convenient visitors from hypothetical planets "out there," who just happen to be humanoid like us, and breathe our air—has run its 70-year-old history. Of course it does explain everything, because there is no limit to miracles if you assume infinity is on your side.

But that's not science.

As precise information appears, as it does in this case, you have to take it into full account, and consider its contradictions.

We do own something like a basket of Easter Eggs: Smarter people than us may be able to take them to the next logical level. Yet all the signs in the world of technology point to the failure of the reverse-engineering programs that have been conducted since the recovery of that "avocado" in San Antonito, one month after the first atomic bomb.

Today, we do have extremely sophisticated systems in orbit and in the outer reaches of the Solar System but the basic contradictions in physics that existed in 1945 have not been resolved, so the technology we use to overcome gravity still rests on the same propulsive concept as the rockets invented by Chinese physicists 5,000 years ago. What we teach in aerospace graduate courses at Stanford and MIT is nothing more than the application of (more or less) controlled chemical reactions to generate thrust, with computer guidance and modern materials.

The question nobody seems to ask is:

"What if those UFO devices had been designed so they could not be reverse-engineered by people with our current level of knowledge and social development? What if their target was at a different level? At a symbolic level, about our relationship to life? At a psychic level, about our relationship to the universe? What if they contained an existential warning?"

^ ^ ^

In a paper Dr. Eric Davis and I published in 2003, titled *"Incommensurability, Orthodoxy and the Physics of High Strangeness: A 6-layer Model for Anomalous Phenomena,"* the last sentence stated: "To test more fully the hypothesis that unidentified aerial phenomena are both physical and psychic in nature, we need much better investigations, a great upgrading of data quality, and a more informed analysis not only of the object being described, but of the impact of the observation on the witnesses and their social environment." (**54**)

The present book attempts to answer that question.

What if the object was a product of a form of information physics (a science in gestation) rather than simply a physical vehicle? What if it was both physical and, for lack of a better word, "psychic"? What was it doing, depositing weird telepathic creatures at an ancient traditional site, one month to the day after mankind's first large-scale, historic liberation of the Atom? (Note)

Was it a direct answer to our discovery of nuclear forces? The hopeful beginning of a dialogue? A message? Was it packaged in such a way that it would trigger the kind of reaction an external actor might need: a forced opening of our minds, the opportunity to set aside our arrogance and to listen to other forms of consciousness, so we could be clearly presented with the flimsy parameters of our survival?

Or a signal, from the point of view of better scientists somewhere, that our survival may not be an inflexible requirement for the universe?

No wonder the so-called "reverse engineering" projects of the United States have been kept so tightly, barely within the law (**55**). Maybe the scientists fear to open the envelope and read out the real message.

Maybe they think it would be too much.

Maybe they're right.

∧ ∧ ∧

Note: Going one step further, a brilliant MIT computer scientist and entrepreneur named Rizwan Virk has suggested that what we experience as the physical universe may be a giant simulation created by far advanced minds, for reasons of their own. In his book *The Simulation Hypothesis*, he presents a well-articulated argument suggesting that humans may be characters in a very advanced kind of videogame. He shows that modern technology is already making it difficult to distinguish between physical reality and projected images impacting on our senses.

EPILOGUE

We were trained not to become a burden on anyone. We were taught not to complain about pain, and so we didn't. However, our friends and family members began to get sick and die. Then there were newborns: loss of eyesight, extra toes or fingers, or lack thereof. Lung cancer. Thyroid. Diabetes and kidney problems. Young people and old people were dying.

Reme Baca, Growing up at Ground Zero, 2011

"Departures from Los Alamos began towards the end of 1945," writes Laura Fermi in her lovely, courageous book *Atoms in the Family* (**8**). She recalls that Enrico and many others like him strongly believed that after four years of classified development, the country had a pressing need for training new generations of atomic scientists in the open, a need as great as or greater than the appetite for more powerful weapons. Once peace was restored, she notes, many of them preferred to go back to teaching and research without the heavy constraints of secrecy.

The Fermi family packed up its New Mexico souvenirs, potteries, photographs and Indian jewels, and headed back for Chicago, where they would help found several Institutes for fundamental research.

Dr. Oppenheimer's last day in Los Alamos was November 16, 1945.

Contrary to a common belief, it wasn't Boston or Silicon Valley that was the cradle of American Science in the twentieth century, but the

Midwest with its deep roots in European culture, its access to scientists who had fled from the Soviet satellites and from traditional Europe, and its constellation of major universities and research centers. (Note)

In March 1946, Enrico Fermi and four others were awarded the Congressional Medal of Merit for their contribution to the development of the atomic bomb.

Laura Fermi had lived through the arc of emigration and war emergencies, raising a family in an unfamiliar culture, moving to an isolated, secret city for four years away from normal society, and re-emerging into a completely changed world where new questions were being asked. A dynamic, intelligent woman, she was mindful of her opportunity to observe those historic changes and to reflect upon them.

As the newspapers began spreading information about the devastations of Hiroshima, the men around her started wondering whether they could really place all the moral responsibility for the massive death toll among innocent Japanese civilians on the Army and the US Government. Both General Eisenhower and Douglas MacArthur had expressed grave moral misgivings to President Truman about using the bomb, but reliable estimates of the loss of life on both sides (actually, three sides since a million Russians had just invaded Manchuria and were ready to devastate the Japanese mainland on Stalin's orders) were equally terrifying.

"There are no unanimous answers to moral problems," Laura reflects in her book. The reactions she observed among the men were varied. Some argued that the quick ending of a horrible, lingering war justified wiping out Hiroshima and Nagasaki. Others, she writes, thought the problem was in the nature of war itself, not in the invention of new weapons. Still others blamed themselves for continuing to build the bomb, once they had demonstrated that it was physically feasible.

Enrico didn't join any of these groups. He didn't think you could ever stop the progress of knowledge itself: it will always find a way to come to the surface, he said. Somebody, sooner or later, would have produced the bomb: it might have fallen into much more evil hands than those of President Harry Truman.

In a letter to Leo Szilard dated July 2, 1945, Dr. Edward Teller had written: "The things we are working on are so terrible that no amount of protesting or fiddling with politics will save our souls."

Note: The first American digital computer was built in Ames, Iowa. Not at MIT or Stanford.

Epilogue

Other members of the scientific team at Los Alamos wanted to go off and hide in shame, and forget they ever worked on the bomb. The Pope disapproved of the weapon, so Catholics were torn by its awful reality. Laura writes there was a sense of indisputable guilt among the scientists on the mesa, more or less consciously; however "it didn't engender demoralization, but hope."

After Hiroshima, the atomic bomb would never be used again, they reasoned. The atomic age was bound to mean international collaboration to share in the benefits of a golden age: "What country would refuse to associate itself with such a program?" The hopeful scientists argued. "Who would walk away from having such enormous electrical power, for pennies?"

^ ^ ^

In October 1945 a group of Los Alamos scientists formed the Association of Los Alamos Scientists, which joined others and grew into the Federation of American Scientists. It issued a public appeal for the creation of a world authority on atomic energy.

Enrico Fermi didn't join: He thought humanity wasn't mature enough for world governance, and that the main problem of war didn't reside in the weapons anyway, but in individual nations' willingness to use them, and to risk suffering the known consequences made very plain in Hiroshima, for everyone to see.

The fact is that we've seen many wars since 1945, including Korea, Vietnam, Afghanistan and multiple conflicts in the Middle East among Israel and Egypt, Iran and Iraq, and more recently the battles of the Arab Spring, without an atomic weapon being used. South America was the theater of more limited, but no less bloody battles, like the Falklands war. In none of these conflicts was the nuclear option employed, nor was it in India against Pakistan, or in Russia against Chechnya. It is the first example of a family of advanced weapons mastered by the military, where those in power have (so far...) refrained from operational deployment.

Only in the Cuban crisis of the Bay of Pigs, in October 1962, did the world come within minutes of a nuclear exchange that could have deteriorated into World War Three, except for the heroic determination of one Soviet naval officer, Commander Vasili Archipov.

How much awareness did Fermi, Oppenheimer and their close colleagues have of the recovery of our strange "avocado" and its even stranger occupants? Was the discovery of that extraordinary object a factor in top secret discussions about the deployment of future bombs?

No word has filtered about that question. But Japan had capitulated, the war was over. The physicists were returning to civilian life. Once again, they were ordinary College professors. They were no longer among the Keepers of the Atomic Secrets. Oppenheimer may not have been told about what Jose Padilla had seen.

New construction had already begun at White Sands, anticipating the future tests of new generations of rockets. In July 1945, *Life Magazine* had published drawings of a manned space station envisioned by the German specialists at Peenemunde. As we've noted before, a group of American scientists was on the way to Europe to assess defeated Germany's rocket development progress *and to collect the equipment*.

Then things happened fast:

In September 1945, White Sands Proving Grounds saw the first launch and development flight of the Army's liquid propulsion WAC Corporal's rocket. It rose higher than four miles.

In October 1945, Secretary of War Patterson approved the plan to bring top German scientists to White Sands under *Project Paperclip* to work on missile development.

In December 1945, over one hundred German rocket scientists and engineers arrived at Fort Bliss, Texas to begin new research that led to a revolution in American ballistic weaponry and to the modern exploration of space.

A new era had begun.

What Jose Padilla and Reme Baca witnessed in August 1945 has lingered in American history as a forgotten mystery.

Everything about the event of the UFO crash at Trinity: its timing, the extraordinary secret surprise it created, the persistent interest of American military Intelligence in its most minute details, the long silence of the witnesses, even the specific place where it crashed and the visions it projected, all that has sunk into the unconscious of a human environment obsessed with progress and power—a human environment thoroughly distracted and confused by the new electronic media that have come to define reality—and replace it for our contemporaries.

∧ ∧ ∧

Epilogue

A physicist friend of mine, now employed on projects involving national security, tells me he's confirmed that statement by Wilbur Smith, asserting that the UFO matter was classified very high, so high that there is no way Paola and I could be briefed by the right specialists on the true nature of the phenomenon.

I am sure he is right about the general scope of the problem, *except that in this particular case we have already been briefed,* very fully and carefully, at length and on site, by three ordinary people with extraordinary courage, two of whom happened to be there, observed the so-called "phenomenon" as it landed, and recovered some of the hard evidence. They were, indeed, the real primary experts because they saw, touched, and actually entered the craft in question, something the so-called "specialists" of today haven't had the privilege to do. So that particular cat, at least, is already out of the bag.

We're still trying to skin that cat.

We don't pretend to have unveiled the full significance of what happened near San Antonito. *Was the object an extraterrestrial craft?* If so, it lacks the accoutrements modern technology would normally associate with such a craft—primary among them, a life support system and space navigation equipment. *Was it a decoy or a warning from another country?* If so, why were the occupants so radically different from known Earth creatures? And what about the projected visions? What about the weird properties of the materials?

All we can say is that this event has taken its place within the tapestry of numerous, well-observed cases with hard, measurable traces and trustworthy witnesses, like Socorro and Valensole, cases that have remained unidentified after all official inquiries by responsible government and scientific agencies were exhausted.

Although pointedly ignored by academic scientists, these incidents are rich with images that have been subliminally injected into our collective psyche and actually amplified by the studious neglect of our leading intellectuals. They still work today inside of us, and they continue to impact human consciousness, through the worldwide media, hinting at cosmic truths.

We ignore them at our peril.

Because of the pressing enigma those undeniable cases present, the story of the Trinity UFO crash remains as a document in human history that we must keep reading, without ever the luxury of firmly turning the page, and closing the book.

ACKNOWLEDGMENTS

This retrospective study of some of the most important UFO cases in recent history would not have been possible without the precious help of many dedicated witnesses, researchers, technical specialists and discreet friends.

Paola and I especially wish to thank the following:

First of all, Mr. Jose Padilla, who generously gave us his time and attention, as well as access to his property for multiple surveys and reviews of the data. He also entrusted us with the unique object he retrieved from inside the craft, which we had the opportunity to analyze. It has now been donated to a University as part of our records of the case, according to his wishes.

Mrs. Sabrina Padilla who completed the picture by giving us her own testimony regarding the metal pieces she handled, and her personal observations at the site a few years after the crash, provided us with precious independent confirmation of the characteristics of the materials.

Mr. Faustino Myers, who kindly contributed his observations of the "angel hair" that had been picked up at the site.

Next, Chuck and Nancy Wade of Gallup, New Mexico, who gave me my first "on the ground" introduction to crash site research in New Mexico over several field surveys, and made available their accumulated knowledge of the recovery of such evidence.

∧ ∧ ∧

Several individuals can only be mentioned *in memoriam*. First among them is my long-term friend Ron Brinkley, who helped me understand the remarkable historical background of his New Mexico family and twice came with me to uncover materials from the 1947 crash site near Datil. He is the one who introduced me to the 1945 San Antonio case shortly before his death.

Sir Arthur Lundahl (1915-1992), recognized as the key organizer of the US post-World War Two image intelligence capability, guided me in the interpretation of some of the unpublished events surrounding the interception of UFOs by the US military. He told me about his extensive personal experiences with the subject inside the Intelligence Community, notably at the time of the extraordinary psychic experiments with "AFFA" and the (still unacknowledged) recovery of unknown material shot out of a flying disk over Washington, DC.

Madame Simonne Servais, a French Ambassador and long-term spokesperson for French Presidents Charles DeGaulle and Georges Pompidou, introduced me to the witnesses of the multiple unidentified (still unreported) cases at Valensole, over several visits at the site.

Mr. Robert Chartrand, a former Naval officer and senior analyst at the Congressional Research Service, tutored me in an understanding of the various structures (such as SAPOC) supervising the complex system of companies controlling the truth about recovered UFO structures. Contrary to all the colorful posturing on television, and to the high visibility of a few "cutouts" and all the hopeful blogging in the Internet about "disclosure," *the real status of the hard data is not about to be published*, possibly for valid reasons. Perhaps our book will change that reality.

Mr. Remigio Baca, one of the key witnesses in the San Antonio crash, agreed to be interviewed by Paola Harris in 2010 and his family generously provided access to the early metallurgical analyses of the objects retrieved from the crash.

∧ ∧ ∧

Many people with special skills in science and in research, and investigators in all walks of life contributed generously with their advice,

Acknowledgments

their skills and their time when they realized that we were serious about pursuing this research to its logical end, and to publish it.

They include:

The Honorable Paul Hellyer, former Defence Minister of Canada, a long-time advisor and confidant of Paola, for providing key material about the early days of this research in North America, contemporary to the events at White Sands.

Mr. Christopher Mellon, former Deputy Assistant Secretary of Defense for Intelligence in the Bill Clinton and George W. Bush administrations, for his support and guidance in the understanding of critical policy decisions.

Professor Garry Nolan, Stanford University School of Medicine, for many stimulating discussions about the phenomenon, and for important advice in proper protocol for chemical analysis and plant pathology study.

Mr. Ray Stanford, for permission to quote his extensive research at the Socorro site at the time of Dr. Hynek's investigation on behalf of Project Blue Book.

Mr. Larry Lemke, for his analysis of the technical parameters of the "bracket" entrusted to us by Mr. Jose Padilla, and for sharing his observations at the site.

Professor Paul Hynek, for many stimulating discussions and recollections of his father. We went to the Socorro site together.

Desta Barnabe, a researcher from Winnipeg in Canada, as well as Mr. Sid Goldberg, Emmy Award winning TV producer from Québec, and Olivia Mackenzie of Boulder, Colorado, for their early work on the Padilla case, in support of Paola Harris' research.

Mrs. Maryann Brown Sperry, who master-minded the final editing of our book for publication.

Sr. Jaime Maussan, professional journalist and TV personality from Mexico, who visited the site with Paola for three days and supervised one of the early metal analyses.

Mr. James Rigney and his associates from Australia, for plant studies and for their exhaustive metal analysis establishing the first complete chemical composition of the "bracket."

Mr. Steve Murillo, of the Paranormal Research Society in Los Angeles, who encouraged my own work and introduced me to Paola Harris in July 2018.

Our common (and uncommon) friend, cinematographer James C. Fox, who surveyed the site at Paola's invitation in 2013.

Flamine de Bonvoisin, who reviewed the research data and helped edit the final draft.

Steve and Kae Geller, who also made insightful comments reflected in our final draft.

Messrs. David Garcia and Tom Hamlin, independent investigators who conducted an early survey of the soil and the vegetation at the crash site in 2015.

Mr. Bill Crowley, who joined our field investigation team in San Antonito in 2019.

Mr. Grant Cameron, for confirming some of the important Wilbert Smith materials.

Ms. Clodagh Domegan, from Vancouver Island, Canada, who transcribed interviews, for all her skillful work on Paola's tape recordings done in the field.

Mr. Dan Farah of Los Angeles and Ms. Yfat Reiss Gendel of New York, for their early encouragements in this project.

Mr. Michael Beltrami, from *Suisse TV*, for his excellent advice in the proper conduct of the field investigations.

Dr. R. Jason B. Reynolds, Botanist at Colorado State University in Fort Collins, for plant pathology analyses based on our collected samples at the crash site.

Mr. Danny Many Horses, for agreeing to speak with me regarding little-known historical information about Apache history in the State of New Mexico, and ancient Apache ceremonies held at the San Antonito site where the crash took place.

And, most importantly,

Everybody at the *Owl Café and Bar*, for good food, good cheer, colorful *ambiance*, and the unique historical perspective!

NOTES AND REFERENCES

1. In the present context "The Plains," or "The Plains of San Agustin" is an American Southwest expression that designates a region in New Mexico south of US Highway 60. The area spans Catron and Socorro Counties, about 50 miles (80 km) west of the town of Socorro and about 25 miles north of Reserve.

   "The Plains" extend roughly northeast-southwest, with a length of about 55 miles (88 km) and a width varying between 5–15 miles (8–24 km). The basin is bounded on the south by the Luera Mountains and Pelona Mountain (outliers of the Black Range); on the west by the Tularosa Mountains; on the north by the Mangas, Crosby, Datil, and Gallinas Mountains; and on the east by the San Mateo Mountains. The Continental Divide lies close to much of the southern and western boundaries of *The Plains*. (From Wikipedia)

2. Dr. Robert Oppenheimer (1904–1967) was a founding father of the American school of theoretical physics that gained world prominence in the 1930s. He was professor of physics at the University of California, Berkeley and the wartime head of the Los Alamos Laboratory in the Manhattan Project that developed the first nuclear weapons. After the war, Oppenheimer became chairman of the Advisory Committee of the Atomic Energy Commission. Oppenheimer's achievements in physics include the Born–Oppenheimer approximation for molecular wave functions, work on the theory of electrons and positrons, the

Oppenheimer–Phillips process in nuclear fusion, and the first prediction of quantum tunneling. After the War, he became director of the Institute for Advanced Study in Princeton, New Jersey.

3. George Adamski (1891-1965), a popular American "Contactee" was born in Prussia in 1891. When he was two-years-old, his family immigrated to the US. After serving in the US Cavalry during the Pancho Villa Expedition (which earned him a place at Arlington Cemetery as a veteran), he married and moved west, doing maintenance work in Yellowstone National Park and working in an Oregon flour mill and a California concrete factory. In the 1920s, Adamski became interested in theosophy and became a minor figure on the California occult scene, teaching "Universal Law" in Laguna Beach, where he founded the "Royal Order of Tibet." In 1944, with funding from Alice K. Wells, Adamski built a wooden observatory with a six-inch telescope at the base of Palomar Mountain. On November 20, 1952, Adamski and his friends were near Desert Center, California, when they purportedly saw an object whose pilot, a Venusian called Orthon, began a friendship that led to alleged trips to other planets and several very popular books, notably *Flying Saucers Have Landed* (co-written with Desmond Leslie) in 1953, *Inside the Space Ships* in 1955, and *Flying Saucers Farewell* in 1961. Several members of the party later confessed they never actually witnessed what they swore to have seen.

   Adamski died in April 1965, generally discredited because his descriptions of the planets he had supposedly visited were proven to be fanciful. However, in a book published in 1983 (*George Adamski, the Untold Story*, with Lou Zinsstag), Timothy Good presents a more nuanced assessment, arguing that the twisted travelogues and questionable photographs were perhaps designed to inspire an early fascination with space, either under the influence of esoteric ideas popular in Southern California, or prompted by "agents" from undisclosed parts of the US military.

4. Frank Scully (1892–1964) was a journalist for the entertainment magazine *Variety*. In 1949, he claimed that dead extraterrestrial beings had been recovered from a flying saucer crash, based on a report by a scientist. His 1950 book *Behind the Flying Saucers* added that there had been two such incidents in Arizona and one in New Mexico, a 1948

incident that involved a saucer 100 feet (30 m) in diameter working on magnetic principles. Scully revealed his two sources as Silas M. Newton and a scientist he called "Dr. Gee." Sixty thousand copies of the book were sold. In 1952 and 1956, *True* magazine published articles by *San Francisco Chronicle* reporter John Philip Cahn that purported to expose Newton and "Dr. Gee" (identified as Leo A. GeBauer) as con artists. Scully briefly revisited the subject in his 1963 book *In Armour Bright*, reiterating his belief in the veracity of a 1948 saucer crash near Aztec, New Mexico.

5. J. Allen Hynek was born in Chicago in 1910, to Czech parents. He received a B.S. from the University of Chicago in 1931 and completed his Ph.D. in astrophysics at Yerkes Observatory in 1935. Joining the Dept of Physics and Astronomy at Ohio State University in 1936, he specialized in stellar evolution and spectroscopic binary stars. During World War II, Hynek helped develop the US Navy's radio proximity fuze at Johns Hopkins.

After the war, Hynek returned to Ohio State. In 1956, he left to join Fred Whipple at the Smithsonian Astrophysical Observatory to direct the tracking of a future American satellite, a project for the International Geophysical Year. In addition to over 200 teams of amateur scientists around the world that were part of Operation Moonwatch, there were also 12 photographic stations. After the satellite program, Hynek went back to teaching as professor and chairman of the astronomy department at Northwestern University in 1960.

In response to numerous reports of "flying saucers", Hynek had been contacted in 1948 to act as a scientific consultant to *Project Sign*, established by the USAF to examine sightings. He concluded that most reports suggested known astronomical objects. In his 1977 book, *The UFO Experience*, Hynek said that he enjoyed his role as a debunker for the Air Force. However, in April 1953, he wrote a report for the *Journal of the Optical Society of America* titled "Unusual Aerial Phenomena," which contained one of his best-known statements: *"Ridicule is not part of the scientific method, and people should not be taught that it is. The steady flow of reports, often made in concert by reliable observers, raises questions of scientific obligation and responsibility. Is there any residue that is worthy of scientific attention? Or, if there isn't, does not an obligation exist to say so to the public—not in words of open ridicule but seriously, to keep faith with the trust the public places in science and scientists?"*

In 1953, Hynek served as an associate of the classified Robertson Panel. It concluded that there was nothing anomalous about UFOs, and that a public relations campaign should be undertaken to debunk the subject and reduce public interest. Hynek remained with *Project Sign* after it became *Project Grudge* and *Project Blue Book* in early 1952. Air Force Captain Edward J. Ruppelt held Hynek in high regard as "one of the most impressive scientists I met while working on the UFO project, and I met a good many. He didn't do two things that some of them did: give you the answer before he knew the question; or immediately begin to expound on his accomplishments in the field of science."

In November 1978, Hynek presented a statement on UFOs before the United Nations General Assembly on behalf of himself, Jacques Vallée, and Claude Poher. Their goal was to initiate a centralized, United Nations authority on UFOs, a recommendation that was never acted upon. Hynek and his wife, Miriam (Curtis) had five children. Dr. Hynek died of a malignant brain tumor, in Scottsdale, Arizona on April 27, 1986.

6. George Hunt Williamson (1926 –1986), aka Michael d'Obrenovic and *Brother Philip*, was a flying saucer contactee, channel, and metaphysical author who came to prominence in the 1950s. A student of archaeology as early as 1946 when he recorded a site in southern Illinois, by 1949, he was a student of anthropology at the University of Arizona and did fieldwork in Lincoln County, New Mexico. In October 1949, he helped found the Yavapai County Archaeological Society in Prescott, Arizona where he lived until 1955. Having read William Dudley Pelley's book *Star Guests* (1950), Williamson worked for Pelley's cult and its publication *Valor*. (See Zirger & Martinelli, *The Incredible Life of George Hunt Williamson*, p. 101). Williamson and co-author Alice Bailey published *The Saucers Speak*, emphasizing supposed short-wave radio contact with saucer pilots after 1952.

7. Dr. Enrico Fermi was born in Rome in 1901, the last of three children. His mother was a remarkable teacher and had a great influence on him. His early research was on relativity and quantum mechanics. He was

the first physicist to split the uranium atom, and received the Nobel Prize in the Fall of 1938 for his work on radioactivity. It was his epoch-making experiments at the University of Rome in 1934 that led directly to the discovery of uranium fission. He took the opportunity of being invited to Stockholm in connection with the Nobel prize to escape to the United States with his wife, at a time when Italy was instituting the dreaded anti-Jewish laws. Enrico Fermi's team built the first nuclear reactor that was first demonstrated in Chicago on December 2, 1942, thereby initiating the Nuclear Age in the opening phase of the Manhattan Project.

Enrico Fermi died of stomach cancer in 1954 in Chicago, age 53.

8. *Atoms in the Family*. University of Chicago Press, 1954, by Laura Fermi. She was born Laura Capon in 1907 in Rome. Her father was an admiral in the Italian Navy. They were an upper-class Jewish family, although they did not practice any religion. As a general science student at Rome University, she met Enrico and married him in 1928. They had two children. Laura Fermi died in 1977, age 70.

9. The excellent US Army brochure *Trinity Site* quotes General Thomas Farrell. He is also prominently mentioned in Chris Wallace's book, *Countdown 1945*.

10. In 1945, the physicist, Leo Szilard, who originally conceived the idea of a nuclear chain reaction, led a group of scientists who wanted to petition President Truman not to use the bomb. His reasoning was that the result would be a nuclear arms race with the Soviets, that a hydrogen bomb would soon be developed, and that World War III and total annihilation would be inevitable. Szilard was despised by General Groves and those who wished to deploy the weapon. They ultimately coerced Oppenheimer into suppressing the petition to President Truman.

11. Ryan S. Wood: *MAJIC Eyes only*. Winfield, KS: Wood Enterprises, 2005. With foreword by Jim Marrs. See pages 52-57 in particular.

12. Reme Baca with Jose Padilla, *Born on the Edge of Ground Zero: Living in the Shadow of Area51*. Privately printed, 2011.

13. Good, Timothy: *Need to Know*. NY: Pegasus, 2007, pp. 27-28

14. Harris, Paola Leopizzi: *Connecting the Dots*. Bloomington, Indiana: Author House, 2008. With foreword by Dr. Leo Sprinkle.

15. Paola Harris' work researching close encounter events and their impact was initially centered on Italian and more generally, European cases rather than US reports. As her work and her international reputation expanded, her interests have included Latin America, particularly Chile and Peru where she made multiple trips and developed long-term relationships with local and national researchers.

16. Project Serpo is an alleged top-secret exchange program between the US and an alien planet by that name. Details have appeared in UFO conspiracy stories, including one incident in 1983 in which a man identifying himself as Air Force Sergeant Richard C. Doty contacted Linda Moulton Howe, offering Air Force records of the exchange for an HBO documentary, only to pull out without providing any evidence. Various Internet versions of the conspiracy theory circulated, involving an Alien who supposedly survived a crash near Roswell, contacted its home planet and was eventually repatriated. This was said to have led to some relationship between the US and a planet of the binary star system Zeta Reticuli, a popular site for fake Aliens in many hoaxes, also mentioned by other self-described "whistleblowers" of uncertain background like Robert Lazar.

The story, which suffers from obvious astronomical flaws and miscalculations in its "scientific proofs," claims that a number of US military personnel (either 12 or 8, depending on which version you believe) visited Mars between 1965 and 1978. A major problem with the yarn, according to Wikipedia, "stems from the lack of veracity of one of its alleged witnesses, Sergeant Richard Doty. Doty has been involved in other alleged UFO-related activities (Majestic 12 and Paul Bennewitz), and thus is a discredited source or a purposeful provider of disinformation." It is worth noting that Doty has also been employed by some UFO research projects in Austin and Las Vegas and has been involved in circulating UFO conspiracy yarns. Most recently, he has appeared in videos created by Dr. Steven Greer.

Notes and References

17. Professor Peter Sturrock, founder of the Plasma Research Institute at Stanford University, has conducted a 40-year study of the explosion (in 1957) of an unknown device over the Brazilian coastal town of Ubatuba. He visited the site, collected some rare samples and pioneered the chemical analysis of such materials not only in terms of elements but also in terms of isotope ratios. He found indications that the recovered samples, essentially composed of extremely pure magnesium, had unusual properties.

    Dr. Sturrock published his findings in a series of peer-reviewed articles, notably:"Composition Analysis of the Brazil Magnesium" in *Journal of Scientific Exploration* (*JSE*) Vol.15, no.1, (2001) pp.69-95, and in "On Events possibly related to the "Brazil Magnesium" *JSE* Vol.8, no.2 (2004) pp.283-291 (with P. Kaufmann).

18. CUFOS and MUFON: Dr. J. A. Hynek was the founder and first head of the Center for UFO Studies (CUFOS). Founded in 1973 in Evanston, Illinois (but later relocated to Chicago), CUFOS advocates for scientific analysis of UFO cases. Its extensive archives include valuable files from civilian research groups such as NICAP, one of the most popular UFO research groups of the 1950s and 1960s.

    The Mutual UFO Network (MUFON) officially began in May 1969 as a spinoff from APRO, the respected *Aerial Phenomena Research Organization* managed by Coral and Jim Lorenzen and based in Arizona. Initially known as the "Midwest UFO Network," MUFON had Dr. Allen Utke, associate professor of chemistry at Wisconsin State University, as its first director. He was replaced by Walt Andrus, who remained in that position until 2000 when he retired and was replaced by John Schuessler, a distinguished aerospace project manager of high reputation, as "International Director." When he retired in November 2006, he was succeeded by James Carrion and later Clifford Clift. David McDonald, a flight school and Air Carrier entrepreneur based in Northern Kentucky, is the current International Director (2020).

19. Vallee, Jacques. "Physical Analyses in ten cases of unexplained aerial objects with material samples." *Journal of Scientific Exploration* (JSE) Vol.12, no.3, pp.359-376 (1998).

20. Moffett, Ben. Article in the *Socorro Mountain Mail*, November 2, 2003.

21. Stanton Terry Friedman (July 29, 1934 - May 13, 2019) was a nuclear physicist, diligent researcher and prolific writer who lived in Fredricton, New Brunswick, Canada. He is best known as the original investigator of the Roswell incident, about which he wrote a series of well-documented popular books. Born in Elizabeth, New Jersey, Mr. Friedman held Bachelor's and Master's degrees in Physics (1956) from the University of Chicago. He worked for many years in classified nuclear propulsion research at such companies as GE, Westinghouse, TRW Systems, Aerojet General Nucleonics and especially at McDonnell Douglas, where he participated (along with Dr. Robert Wood) in classified investigations of UFO sightings.

22. Dr. Oppenheimer and the AEC Hearings: The way Dr. Oppenheimer was treated by the US Security establishment has left its indelible mark upon the academic world and has discouraged many scientists and intellectuals from dealing with US classified agencies which they see as increasingly clannish and often corrupt. In the field of UFO research, obvious hoaxes like the Serpo hoax and the convoluted series of MJ-12 "revelations" (there was a real MJ-12 at one time, but it wasn't dealing with the same topics) have further discouraged many scientists from risking their otherwise impeccable careers "chasing shadows."

23. The "Marconi Tower" was located very close to the entrance gate to the Padilla property, at approximately 5,000 feet altitude. Resting on four legs anchored in cement, it rose to a height of 75 feet.

24. Members of several families, not surprisingly, were neighbors or relatives within the small towns involved in both the San Antonio events and the Socorro landing.

25. Remigio Baca died on June 12, 2013, of complications of diabetes. He was survived by his wife Virginia, two children and seven grandchildren. He was 75-years-old.

    "Reme" had moved to Washington State in 1954 and enlisted in the Marines after high school graduation, becoming a master mechanic. After settling in Tacoma, he held a number of Governor-appointed positions in support of the Mexican-American community in the

State of Washington for the next 50 years. He retired in Gig Harbor, remaining active in the Democratic Party.

26. Bartimus, Tad, and McCartney, Scott: *Trinity's Children. Living along America's Nuclear Highway.* Harcourt Brace Jovanovitch, 1991 (Political Science). Reviewers describe the book as an "Eye-opening and evenhanded report by two AP journalists on the history of the nuclear weapons industry in the Southwest and its effects on its employees and neighbors."

    They add: "Bartimus and McCartney (...) concentrate on the industry's effects on the environment and the neighbors who share air, land, and water with the Rocky Flats nuclear weapons plant, the nuclear weapons storage facility at Kirtland Air Force Base in Albuquerque, Los Alamos National Laboratory, MX missile silos, and other facilities. Ordering their survey geographically from Trinity Site to the missile silos of Wyoming, the authors offer tales of lost ranch land, displaced citizens, poisoned employees, and terrified mothers, but are careful to include thought-provoking responses from within the nuclear industry (scientists, engineers, and commanding officers) as well."

27. "An investigation into an unknown metal sample taken from San Antonio, New Mexico," by J. Nguyen, C.E. Lal and J.W. Auchettl. Mulgrave, Victoria, Australia: Phenomena Research Australia, privately printed.

28. Similar Aermotor products can be found at www.aermotor.com notably a bracket-like device (part A585) called a "Windmill tailbone casting." Re. Dean Bennett.

29. Long after the events at Ground Zero, Mr. William Brophy, Sr. (an experienced pilot based in Alamogordo) told his son Bill that the crew of a B-25 on a training mission over Walnut Creek had reported seeing some smoke that day.

    They thought a plane might have hit the tower and crashed. According to Bill Brophy, his father's commanding officer, General Maurice Arthur Preston, put William J. Brophy, Sr. in charge of the investigation, which reportedly led to Brophy seeing the crashed object and the "two little Indian boys" on horseback. He was in charge until the next day, when Army Colonel Turner took over responsibility for the recovery of the craft. William Brophy Sr. may have evacuated the craft's occupants, partially camouflaged it, and cleaned up the crash site on the second day.

30. Jeff Kripal and Whitley Strieber, *The Supernatural*. p.128. NY: Tarcher Perigee, 2016.

31. The scientists themselves, under Oppenheimer, may not have been told about the UFO crash. What makes us wonder about this is the fact that Dr. Phillip Morrison, one of the elite physicists in the group, later commented about UFOs in the pages of the prestigious *Science Magazine*, arguing that reported "landings" of such objects could not be attributed to Alien visitors because the supposed craft did not leave the damage or chemical residues one would expect from an interstellar rocket taking off... If his letter was not written as a disinformation plot at the instigation of the military, it shows a surprising lack of technical acumen in a scientist who contributed to building the first atom bomb.

32. Major Quintanilla's staff was not always so relaxed. While it took the Major three days to fully react to the Socorro event, Hynek was impressed when the FTD team rushed to Michigan one morning, proceeded to cordon off a wooded zone where an object had reportedly crashed, and ordered dozens of soldiers to guard the area and explore it for debris. In that particular case, the object in question was a Soviet satellite whose re-entry had been expected by the orbit trackers, and the technicians at the Foreign Technology Division were eager to get their hands on the Russian material. The UFO interest at FTD was only a public relations job, but the same team had more immediate responsibilities, which ufologists and the media never took into serious consideration.

33. See Ray Stanford's book, *Socorro Saucer in a Pentagon Pantry*. Austin, Texas: Blue Apple Books, 1976. Officer Lonnie Zamora gave multiple testimonies, including in lengthy interrogation before hostile military and law enforcement investigators, in the three days following his sighting. He was even accused by some of his own colleagues of having faked the incident to attract tourists to his town!

34. Hynek, Dr. J. Allen: *The UFO Experience*. Chicago: Henry Regnery, 1972, especially pages 144 and 145 for the Socorro case.

35. Bill Powers: *The Landing at Socorro – New Light on a Classic Case*. FSR Special Issue "The Humanoids", Oct-Nov. 1966, pp. 47-51 (Charles Bowen, editor).

36. NICAP, the "National Investigation Committee on Aerial Phenomenan" was founded in October 1956, by inventor Thomas Townsend Brown. The board included Marines Major Donald Keyhoe and former chief of the Navy's guided missile program, rear-admiral Delmer S. Fahrney, who convened a press conference on January 16, 1957, where he announced that UFOs were under intelligent control, but that they were of neither American nor Soviet origin. In April 1957, Fahrney resigned, citing personal issues.

Keyhoe became director and established a newsletter, *The U.F.O. Investigator*. He brought his Naval Academy classmate Vice-admiral Roscoe H. Hillenkoetter, first head of the CIA, as Chairman. Another important member was Gen. Albert Coady Wedemeyer.

The organization had chapters and local associates throughout the US. Many members were amateurs, but a considerable percentage were professionals, including journalists, military personnel, scientists and physicians. They would eventually compile a significant number of case files and field investigations. By 1958, NICAP had grown to over 5,000 members. The 1960s found much of the American public keenly interested in UFOs, and membership crested at 14,000, greatly improving the group's wobbly finances.

Hillenkoetter left the board in Feb 1962 and was replaced by a former covert CIA high official, Joseph Bryan III, the CIA's first Chief of Political & Psychological Warfare. Bryan never disclosed his CIA background.

In 1964, NICAP published *The UFO Evidence*, edited by Richard H. Hall, a summary of hundreds of unexplained reports studied by NICAP investigators through 1963. The book is still considered an invaluable reference. Following the 1968 Condon report at the University of Colorado (which concluded there was nothing extraordinary about UFOs), public interest in the subject abated, and NICAP's membership dropped.

When Keyhoe resigned in 1969, Joseph Bryan took over and disbanded the affiliate groups. Afterwards, John L. Acuff was named director while the organization became paralyzed by infighting, including charges that the CIA had infiltrated it. NICAP was dissolved in 1980. The files were eventually purchased by CUFOS for $5,000. (Summarized from a Wikipedia entry-JV).

37. Ray Stanford, op.cit., quoted with permission.

38. Dr. Henry E. Frankel, who died on September 23, 1989, was a distinguished NASA scientist. The founder of the Materials R&D branch at Goddard from 1960 to 1970, and chief of the engineering division from 1970 to 1974, he was the ideal expert to consider the evidence recovered by Ray Stanford. At the time of his death, he was Chief Scientist at Orbital Research, Inc. (Source: Goddard Retirees and Alumni Association).

39. Mr. Thomas P. Sciacca's statement was put in writing by NASA over his signature in an official letter to NICAP. Reference: File 3875, 623.8(64)-40/TS/cmd.

40. CNES is the *Centre National d'Etudes Spatiales*, the space agency of the French government, based in Paris and Toulouse with a very extensive launch base in Kourou, French Guiana, close to the equator. Through ownership of shares in leading industrial companies, CNES is well-capitalized, the third most important space organization behind NASA and Roscosmos.

41. That significant physiological "detail" from Masse's testimony describing his "paralysis" was taken very seriously in France. It later spurred intense research at the French National Center for Space Studies in Paris and Toulouse, and within French military laboratories.

42. Aimé Michel, "The Valensole Affair," in FSR XI, no.6, (1965) pp.7-9 and "Valensole, Further Details," XII, no.3, pp.24-25.

43. Dr.Bernard E. Finch, M.D., "Comments on the Valensole Affair," FSR XII no.1, pp.14-15.

44. Dr. Hemplemann: "We got away with it." See Reme Baca with Jose Padilla, *Born on the Edge of Ground Zero*. Privately printed, 2011. pp.88-89.

45. *Project Magnet* was an unidentified flying object study program officially established by Transport Canada in December 1950 under the direction of Wilbert Brockhouse Smith, a senior radio engineer.

    The Canadian equivalent of *Project Blue Book*, it was formally active until mid-1954 and informally active (without government funding) until Smith's death in 1962. Smith eventually concluded that UFOs were probably extraterrestrial in origin and likely operated by manipulation of magnetism. In October 1952, Smith had set up an observatory at Shirley's Bay outside Ottawa to study reports of UFO sightings, believing that UFOs would have physical characteristics that could be measured. Numerous sighting reports were investigated by *Project Magnet* until its closure in 1954.

46. *Le Parisien*, Paris Métro édition, 23 Nov.1966, quoting a paper by well-known American journalist Frank Edward.

47. George Popovitch interview with Wilbert Smith: Mr. Francis Ridge (ex-NICAP) produced the transcript of a face-to-face interview with Mr. Smith by George Popovitch (of Akron, Ohio) and C.W. Fitch, of Cleveland. Web reference: Frank Edwards, *Flying Saucers, serious business*, Lyle Stuart 1966 pocketbook edition, section "Pick Up the Pieces," pp.47-50. The detailed transcript is available at: www.nicap.org/reports/520723washington_transcripts.htm

    Note that Donald Keyhoe briefly mentions the case in *The Flying Saucer Conspiracy* (NY: Holt hardcover, 1955) page 272.

48. Sir Arthur Lundahl, private communication. The discussion with me was at his home in Bethesda, Maryland, on September 14, 1989, along with Mr. Robert Chartrand of the Congressional Research Service.

49. NIDS, the National Institute for Discovery Science (NIDS) was a privately financed research organization based in Las Vegas, Nevada, USA, operated from 1995 to 2004.

It was founded by real-estate developer Robert Bigelow, who set it up to research and advance serious study of various fringe science topics, most notably ufology and survival of bodily death. Deputy Administrator Dr. Colm Kelleher, a biologist, was quoted as saying the organization was not designed to study UFOs only: "We don't study Aliens, we study anomalies. They're the same thing in a lot of people's minds, but not in our minds."

The permanent scientific staff also included Dr. Eric Davis, a theoretical physicist, and Dr. George Onet, a veterinarian who led research on cattle mutilations. "The science board of NIDS included two astronauts who had worked on the Moon (Ed Mitchell and Harrison Schmitt) as well as aerospace physicians, space technologist John Schuessler, physicists from Los Alamos, and statistician Dr. Jessica Utts. Also on the board were Dr. Christopher (Kit) Green, Dr. Harold Puthoff, and computer scientist Dr. Jacques Vallee."

NIDS served as a way to channel private funds into the scientific study of paranormal phenomena. It performed research in the area of cattle mutilation and black triangle reports. NIDS bought Skinwalker Ranch after TV journalist George Knapp wrote about it in 1996, and Deputy Administrator Colm Kelleher led the investigation for a number of years. A hotline was established in 1999 to receive reports of odd occurrences. Over 5,000 calls and e-mails were received by the organization; many were explained as missile test launches and meteors. NIDS was disbanded in October 2004 when Robert Bigelow decided to focus all his future investments on space station development.

50. Notes, October 16, 2002. A copy of Eric Davis' 15-page notes from the meeting was among the private papers of astronaut Edgar Mitchell. After the information was acquired by Australian researcher James Rigney, it evidently ended up on the web following Captain Mitchell's death. It became known as the basis for the "Core Secret."

51. The narrative was relayed by researcher Richard Dolan in his *YouTube* overview exposing the Eric Davis memo. The allegations have been formally denied by retired Vice-Admiral Thomas Wilson. On June 15, 2020 he was quoted by reporter Billy Cox as saying "I wouldn't know Eric Davis if he walked in right now," although he mentioned they might have been in the same meeting at some point. The transcript

of the Billy Cox interview includes the statement that "Former Los Alamos National Laboratory scientist Oke Shannon, former Principal Deputy Under-Secretary of Defense for Acquisition, Technology and Logistics Noel Longuemare, former Vice Chair of the Joint Chiefs of Staff and USAF Gen. Joe Ralston, and former Under Secretary of Defense for Acquisition, Technology and Logistics Dr. Paul Kaminsky, have all stated they have no knowledge of the 'Core Secrets' version of events."

The source of the story was Apollo astronaut Edger Mitchell, who recounted his own meeting with Admiral Wilson for CNN audience on *Larry King Live*, July 4, 2008.

52. Annie Jacobsen, *Area 51*. New York: Little Brown – Bay Back Books, 2011. Especially pages 368-369.

53. EG&G was founded as a consulting firm in 1931 by MIT professor Harold Edgerton (a pioneer of high-speed photography) and his graduate student Kenneth Germeshausen, joined by Herbert E. Grier in 1934. Bernard «Barney» O'Keefe became the fourth member of their fledgling technology group, whose high-speed photography was used to image implosion tests during the Manhattan Project. During the 1950s and 1960s, EG&G was involved in nuclear tests as a major contractor for the AEC at the Nevada Test Site for weapons development, and at Nellis AFB. EG&G has shared operational responsibility for the NTS with Livermore Labs, Raytheon Services Nevada, Reynolds Electrical and Engineering (REECO) and others. Subsequently EG&G expanded its range of services, providing facilities management, technical services, security, and pilot training for the military and others.

EG&G builds sensing, detection and imaging products including night vision equipment, sensors for detection of nuclear material and chemical and biological weapons agents, and "a variety of acoustic sensors." The company also supplies microwave and electronic components to the government, security systems, and systems for electronic warfare and mine countermeasures. During the 1970s and 1980s, led by O'Keefe, it diversified by acquisition into the fields of paper making, instrumentation for scientific, marine, environmental and geophysical uses, automotive testing, fans and blowers, frequency control devices and other components including BBD and CCD technology via their Reticon division.

In the late 1980s and early 1990s most of these divisions were sold, and on May 28, 1999, the non-government side of EG&G purchased the Analytical Instruments Division of PerkinElmer for $425 million, also assuming the PerkinElmer name (NYSE: PKI). Based in Wellesley, Massachusetts, it makes products for automotive, medical, and Aerospace applications.

From 1999 until 2001, EG&G was wholly owned by The Carlyle Group and in August 2002, the defense-and-services sector was acquired by defense technical services giant URS Corporation in Gaithersburg, MD employing over 11,000 people. (During its heyday in the 1980s, EG&G had about 35,000 employees). In December 2009, CEO Martin Koffel indicated that the EG&G Division would become "URS Federal Services."

In 2014, URS was acquired by AECOM. In the present context it should be noted that EG&G's "Special Projects" division was the notable operator of the Janet Terminal at McCarran International Airport in Las Vegas, NV, a service used to transport employees to classified government locations in Nevada and California. EG&G also had a joint venture with Raytheon Technical Services, creating JT3 in 2000 to operate the Joint Range Technical Services contract. In the UFO literature the name of EG&G has been linked to a number of wild claims, included those made by Robert Lazar regarding his supposed contact with live Aliens at the highly-classified site.

54. *"Incommensurability, Orthodoxy and the Physics of High Strangeness: A 6-layer Model for Anomalous Phenomena,"* by Vallee & Davis. This paper was presented at the International Forum on "Science, Religion and Consciousness" at the University Fernando Pessoa, Porto (Portugal) on October 24, 2003. The last sentence in the paper under *Conclusion* stated: "To test more fully the hypothesis that UAP phenomena are both physical and psychic in nature, we need much better investigations, a great upgrading of data quality, and a more informed analysis not only of the object being described, but of the impact of the observation on the witnesses and their social environment." The full text of the paper is available on the web on Jacques Vallee's website, www.jacquesvallee.net

55. Advanced scientific projects, secret or not, can be kept private for a very long time, beyond the scrutiny of most government reporting. The case of the Battelle Memorial Institute in Columbus, Ohio, is one example among several. The Institute, with respected expertise in metallurgy, was a classified contractor to Project Blue Book in the mid-1950s. They did the first extensive statistical study of UFO sightings reported to the Air force. At the time of the CIA-sponsored "Robertson Panel" review of Project Blue Book (1954), they issued a classified letter requesting postponement of the meeting for various scientific reasons. I discovered that letter by accident in the files of Dr. Hynek and it was subsequently declassified and published, against much controversy within the UFO amateur community trying to discredit the revelation, thus providing rare insight into what was going on behind the scenes.

Battelle has maintained its great reputation in metallurgy research. *Business Wire* reported on November 10, 2020 that the Institute had been awarded a 7-year, $46.3M contract to study the manufacture of materials for "extreme hypersonic environments." On July 24, 2020 the Institute created an alliance of industry partners to design a nuclear power system for the moon (*Idaho Nat'l Lab press release*). Organized in 1929 as a nonprofit charitable Trust, the unique structure of the Institute places it beyond ordinary audits and makes it immune from FOIA requests and most oversight.

BOOKS AND MONOGRAPHS

The literature concerned with reports of UFO "crashes" and physical evidence, and related historical and political events, represents hundreds of books and monographs, notes and files scattered among the many groups interested by the phenomenon. When foreign titles are taken into account, the number of books is in thousands.

We have only listed here some of the most relevant titles, particularly those contemporary documents we have consulted in our own research about the nature and context of the reported phenomena.

The primary reference to the events described here is a monograph by Reme Baca and Jose Padilla, entitled: *Born on the Edge of Ground Zero: Living in the Shadow of Area 51.* Privately printed, 2011.

Note that the title is a bit strange, since Area 51 is far away in Nevada and did not even exist at the time when atomic experiments were going on at "Ground Zero" in the State of New Mexico.

Few of the following books mention the 1945 San Antonio crash. It essentially remained unknown until a web genealogy search re-established communication between Padilla and Baca after 2003. But these books and other documents were important in analyzing the relevant phenomena, their history, and their aftermath.

As we completed the final draft of this book, two important volumes were published in the US about the Manhattan Project and the defeat of Japan. They were David Barrett's *140 Days to Hiroshima* and Chris Wallace's *Countdown 1945*. Neither mentioned UFOs, but they were useful to us for fact-checking about the operations in New Mexico that particular summer.

Barrett, David Dean: *140 Days to Hiroshima*: The story of Japan's last chance to avoid Armageddon. New York: Diversion Books, April 2020.

Bartimus, Tad and McCartney, Scott: *Trinity's Children. Living along America's Nuclear Highway.* Harcourt Brace Jovanovitch, 1991 (Political Science).

Basterfield, Keith: *Catalogue of Material Fragments*, etc. version 3.3 (privately printed).

Brown, Anthony Cave: *Bodyguard of Lies.* In two volumes. Harper & Row, 1975. Hardcover.

Callaghan & Co. *Relative Abundances of Naturally-occurring Isotopes.* Privately printed, Callaghan & Co. San Jose, California.

Corso, Philip J. with William J. Birnes. *The Day after Roswell.* NY: Pocket Books, 1997. (Foreword by Senator Strom Thurmond).

Davis, Eric and Vallee, Jacques: "*Incommensurability, Orthodoxy and the Physics of High Strangeness: A 6-layer Model for Anomalous Phenomena.*" Presented at the Forum on "Science, Religion and Consciousness" at the University Fernando Pessoa, Porto (Portugal). Published October 24, 2003.

Dolan, Richard M. *UFOs and the National Security State, an Unclassified History.* (Foreword by Jacques Vallee). Volume One: 1941-1973. Keyhole Publishing Company, 2000.

Edwards, Frank. *Flying Saucers – Serious Business.* NY: Lyle-Stuart Books, 1966.

Fawcett, Larry and Barry Greenwood. *Clear Intent: The Government Cover-Up of the UFO Experience.* NY: Prentice-Hall, 1984.

Fermi, Laura. *Atoms in the Family – My life with Enrico Fermi* Chicago: University of Chicago Press, 1954. (Also in paperback, 1961) French edition: *Atomes en Famille.* Paperback. Paris: Gallimard, 1955.

Books and Monographs

Ford, Daniel. *The Button*. NY: Simon & Schuster, 1985.

Friedman, Stanton, and Don Berliner. *Crash at Corona*. NY: Marlowe, 1992.

Fuller, John G. *The Interrupted Journey*. London: Souvenir Press, Ltd., 1980.

Good, Timothy. *Need to Know*. NY: Pegasus, 2007.

Good, Timothy. *Above Top Secret*. Sidgwick & Jackson, 1987.

Groves, Leslie R. *Diary of Lt. General Leslie R. Groves*. US National Archives. Record Group 200, National Archives Gift Collection, Microfilm roll no. 2.

Hacker, Bart. *The Dragon's Tail: Radiation Safety in the Manhattan Project 1942-1946* Berkeley: University of California Press, 1987. See esp. pp.104-105.

Haines, Gerald. "CIA's Role in the Study of UFOs 1947-90." *Studies in Intelligence*, Vol.1, No.1, Central Intelligence Agency, 1997. See also www.cia.gov Note: this author should not be confused with Dr. Richard Haines, from NASA (Ret'd)

Hanks, Micah. *The UFO Singularity*. Pompton Press, NJ.: The Career Press, 2013.

Hansen. *Constitution of Binary Alloys*. Standard science reference.

Harris, Paola Leopizzi. *Connecting the Dots: Making sense of the UFO Phenomena*. Bloomington, Indiana: Authorhouse, 2008. (Foreword by Dr. Leo Sprinkle).

Hellyer, Paul. *Light at the End of the Tunnel*. 2010.

Hynek, J. Allen. *The UFO Experience*. Chicago: Henry Regnery, 1972, especially pages 144 and 145 for the Socorro case.

Jacobsen, Annie. *Area 51*. New York: Little Brown – Bay Back Books, 2011.

Jung, Carl G. *Flying Saucers – A Modern Myth*. In *Complete Works*, Princeton University Press.

Kripal, Jeff and Whitley Strieber. *The Supernatural. A New Vision of the Unexplained* NY: Tarcher-Penguin, 2016.

Lim, Xiaozhi. *Metal Mixology. Nature Magazine* Vol. 333, pp. 306-307, May 19, 2016.

Marcel, Jesse (Jr.) *The Roswell Legacy.* Helena, Montana: Big Sky Press, 2007. (Foreword by Stanton Friedman)

Nolan, Garry, Vallée, Jacques and Faggin, Federico: Towards Multi-disciplinary SETI Research. Presented as part of a SETI series of papers on future research directions. Mountain View, CA. 2018.

Nolan, Garry and Vallée, Jacques F. What do we know about the Material Composition of UFOs? Paris: Institut Métapsychique, 2016.

Pasulka, Diana W. *American Cosmic: UFOs, Religion, Technology.* NY-London: Oxford 2019.

Phillips, Ted. Physical Traces Associated with UFO Sightings: A Preliminary Catalog (Evanston, IL: Center for UFO Studies, 1981).

Poher, Claude and Vallée, Jacques. *Basic Patterns in UFO Observations.* AIAA 13th Aerospace Sciences Meeting, Paper no.75-42 Pasadena, CA. 20-22 January 1975.

Redfern, Nick. *Final Events.* San Antonio, Texas: Anomalist Books, 2010.

Redfern, Nick. *Body Snatchers in the Desert: The Horrible Truth at the Heart of the Roswell Story.* NY: Paraview Pocket, 2005.

Ryan, Craig. *The Pre-Astronauts: Manned Ballooning on the Threshold of Space.* Annapolis: Naval Institute Press, 1995.

Schuessler, John. "Recognizing 'Alien' Metals." *Omni Magazine* Nov.1979, pp.130-132.

Scully, Frank. *Behind the Flying Saucers.* NY: Henry Holt, 1950.

Smith, Wilbert B. *The New Science.* Victoria, Canada: Fenn-Graphic Publishing, 1964.

Southall, H.L. and Iberly, C.E.: System Considerations for Airborne, high-power Superconducting Generators. *IEEE Mag.* 15, 1, 711.

Stanford, Ray. *Socorro 'Saucer' in a Pentagon Pantry.* Austin, TX: Blue Apple Books, 1976.

Strugatsky, Arkady and Boris. *Roadside Picnic.* NY: Pocket Books/Kangaroo, 1978.

Stringfield, Leonard H. *Situation Red: The UFO Siege*. Sphere Books, Ltd., 1977.

Stringfield, Leonard H. *UFO Crash/Retrievals: The Inner Sanctum*. Status Report VI. Privately printed, 1991.

Swickard, Michael. *The Real Contamination of New Mexico*. NMPolitics.net, 2010.

Torres, Noe and Ruben Uriarte. *The Other Roswell*. www.Roswellbooks.com, 2008.

US Army, *Trinity Site*, official brochure available through the Public Affairs Office, Building 1782, White Sands Missile Range, NM. 88002-5047. (Highly recommended.)

Vallée, Jacques F. Physical Analyses in ten cases of unexplained aerial objects with material samples. *Journal of Scientific Exploration* (JSE) Vol.12, no.3, pp.359-376 (1998).

Vallée, Jacques F. *Forbidden Science*: Journals, Vol 1 to 4. NY: Anomalist Press, 2007-2020. Also available from Amazon.com online or in print.

Vallée, Jacques F. "The Psycho-Physical Nature of UFO Reality: A Speculative Framework." AIAA Thesis-Antithesis Conference Proceedings. Los Angeles, 1975, pp.19-21.

Vesco, Renato and David Hatcher Childress. *Man-made UFOs, 1944-1994, 50 Years of Suppression*. Stelle, Illinois: AUP Publishers Network, 1994.

Virk, Rizwan: *The Simulation Hypothesis*. Bayview Books, 2019.

Wallace, Chris, and Weiss, Mitch: *Countdown 1945*. NY: Simon & Schuster/Avid Reader Press, June 2020.

Wood, Ryan S. *MAJIC Eyes only: Earth's Encounters with Extra-terrestrial Technology*. Broomfield, CO: Wood Enterprises, 2005. (Foreword by Jim Marrs).

PEOPLE INDEX

Adamski, George: xiii, xvi, 320
Alliare: 12
Anasazi (Indian Tribe): 12
Anaya, Pedro (see also "Pedro"): 145
Apache (Indian Tribe): 11-13, 18, 63, 142
Apodaca, Eddie: 25, 27, 88, 104, 110, 144, 158
Archipov, Cmdr. Vassili (USSR): 311
Arnold, Kenneth: x, xiv, xvii, 33, 41, 117, 147, 214

Baca, Alejandro: 145
Baca, Cristobal: 12
Baca, Remigio ("Reme"): 11, 12, 14-17, 19-54, 73, 85-103, 112, 114, 117, 134-137, 139, 141, 146, 154, 159, 220, 225, 227, 233, 243, 256, 264, 273, 293, 296, 304, 309, 312
Baca, Virginia ("Ginnie"): 45, 46, 86, 90, 92, 93, 227, 228
Barbosa, Dr. Luisa: 119
Barnabe, Desta: 317
Barrett, David Dean: 115, 116
Bartimus, Tad: 113
Beltrami, Michael: 318
Bergier, Jacques: 7
Bethe, Prof. Hans: 60
Bohr, Prof. Niels: 205
Born, in "Born-Oppenheimer approximation": 319
Bourret, Jean-Claude: 218
Brinkley, Ron: ix, xv (fig.2), xix, 53, 60, 65, 316
Brophy, Bill: 18 (Fig. 8), 20, 33, 110, 137, 143-144, 327, 328

Brown, Anthony Cave: 181
Burns, Howard (FBI): 189, 190
Burroughs, Edgar Rice: 146

Cameron, Grant: 298
Case, William: 116, 117
Chartrand, Robert: 316
Chavez (Officer Sam): 189, 190, 195
Childress, David Hatcher: 158
Chirikahua (Indian Tribe): 13
Churchill, Winston: xiii
Clark, Jerome: 195, 207
Condon, Dr. Ed: 120, 197
Conners, William (USAF): 190
Corso (Colonel Philip): 50, 51, 53, 233, 274, 295, 300, 354
Crowley, Bill: 67, 75, 318
Dahl, Harold: 117
Davis, Dr. Eric: 300, 301, 307, 332
Dolan, Richard: 301, 332
Domegan, Clodagh: 318

Easton, James: 205, 206, 207, 208
Edwards, Frank: 118, 331
Eisenhower, Gal. Dwight: 310
Emperor (of Japan): xiii, 5

Farrell, Brig.Gal Thomas: 4, 9, 57, 69, 323
Fermi, Enrico: 1, 2, 4, 5, 6, 60, 73, 114, 271, 287, 310, 311, 322, 323
Fermi, Laura: 1, 2, 4, 287-291, 296, 309, 310, 323, 338
Finch, Dr. Bernard E., MD: 223, 224
Fitch, C. W.: 118, 298, 331
Fontes, Dr. Olavo, MD: 119

Fort, Charles: 146, 205
Fox, James: 16, 77, 79, 136, 318
Frankel, Dr. Henry (NASA-Goddard): 210, 211, 330
Friedman, Stanton: xvi, 88, 326, 339, 340

Garcia, David: 77, 318
Gebauer, Leo A. ("Dr.Gee"):321
Geller, Steve & Kae: 318
Goldberg, Sid: 139, 317
Good, Timothy: 15, 21, 44, 136, 139, 320, 324, 339
Gore, Congressman Al: 298
Greninger: 47
Grinder, Opal: 194-6
Gurney, N. Joseph: 116

Hall, Richard (NICAP): 210, 329
Hamlin, Tom: 77, 78, 318
Hanlon, Don: 116
Harbinson, Tom, 78
Harbinson, W.A., 159
Harris, Paola: throughout the book
Hellyer, Paul: 50, 51 (fig.10), 317
Hemplemann, Dr. Louis: 1, 296
Hermes, Robert: 72
Hilton, Conrad: 12
Hitler, Adolf: 181
Hohokam (IndianTribe):12
Holder, Richard T. (USAF Captain): 189, 190, 196, 204, 208
Hynek, Dr. J. Allen: x, xv, xvii, 53, 75, 79, 147, 177, 179, 180, 182, 185-188, 193, 195-197, 200-202, 206-207 (Fig.28), 209, 214-215, 228, 230, 234, 299, 317, 321, 325, 328, 329, 335
Hynek, Paul: 97, 228, 317

Hynek, Miriam ("Mimi"): 185, 322
Jacobsen, Annie: 303, 333
Jahn, Professor Robert: 101
Johnson, President Lyndon B.: 263
Jordan, Officer Ted: 202, 204, 207, 303

Kennedy, Pres. John F.: 90
Kies, Paul: 196
Kilou: 219, 222
Kimbler, Major Frank: 62
Klass, Philip: 179
Knowles, Admiral Herbert Bains: 118
Knuth, Kevin H.: 270
Korzybski, Alfred: 160, 169
Kratzer, Larry: 196
Kripal, Professor Jeffrey: 149

Lemke, Larry: 122, 317
Lopez, Nep: 196
Lorenzen, Coral: 120, 325
Lundahl, Sir Arthur: 294, 298, 299, 331

Mack, Dr. John: 148
Mackenzie, Olivia: 317
Maffei, Dr. Risvaldo: 119
Many Horses, Danny: 12, 13, 318
Marconi (tower): 62
Marshall, General George: 115
Martinez, Mike (FBI): 189
Masse, Maurice: 214-224
Maussan, Jaime: 16, 136, 317
MacArthur, General Douglas: 310
McGarey (NICAP): 210
McCartney: 113, 327, 338
McDonald, Dr. James: 179
Mengele, Nazi Dr.: 303, 304
Menzel, Prof. Donald: 299

People Index

Michel, Aimé: 223, 330
Miller, Lt-Commander Willard: 300
Mitchell, Edgar (Captain): 300, 332, 333
Moffett, Ben: 86, 87, 142, 325
Mogollon (Indian tribe): 12, 14
Moody (Sergeant): 180, 190, 195
Mooradian: 47
Morrisson, Dr. Philip: 60, 328
Murillo, Steve: 53, 59, 318
Myers, Mrs. Esther: 277, 278, 282-288
Myers, Faustino: 273, 274-288
Myers, Frederic: 149

Newton, Silas: 321
Nolan, Dr. Garry: 80, 122, 317, 340

Olander, Arne: 47
Oliva (Commander): 215
Oppenheimer, Dr. Robert: xii, 5, 7, 60, 69, 89, 92, 114, 233, 271, 309, 311, 312, 319
Oppenheimer-Phillips process: 320
Orthon: 320

Padilla, Faustino: 11, 14, 16, 17, 19, 24-27, 29, 30, 37, 141, 143-145, 158, 229-230, 235-236, 238-239, 246-247, 253, 258, 260-261, 263, 273
Padilla, Inez: 14
Padilla, Jose: throughout the book.
Padilla, Sabrina: 232-233, 235-247, 251-268, 270, 271 (Fig.30), 273-288
Padilla, Trini: 14
Palmer, Ray: xvii, 118
Patterson (Secretary of War): 312
Pedro the sheepherder (Anaya): 145

Perry, William (Secretary of Defense): 300
Pieris, Genia: 4,
Poher, Dr. Claude: 132
Popovitch, George: 118, 298, 331
Powell, Robert M.: 270
Powers, William T.: 186, 199
Preston, General Maurice: 33, 327
Puthoff, Dr. Harold ("Hal"): 121, 332

Quintanilla, Hector (Captain): 177, 180, 182, 185, 187, 193, 195, 328

Reali, Peter A.: 270
Reeves family (neighbors): 258
Reynolds, Dr. Jason B.: 78, 318
Rigney, James: 78, 139, 317
Rumsfeld, Donald (Defense Secretary): 299
Ryan, Craig: 136, 341

Sagan, Dr. Carl: 179
Santos, Bernabe Hernandez: 65
Saunders, Dr. David: 179
Schalin, Sven: 119
Schuhart, Redrick: 250
Sciacca, Thomas P., Jr.: 211, 330
Scoles, Sarah: 131
Scully, Frank: 320, 321
Ségré, Dr. Emilio: 1
Servais, Mme Simonne: 316
Sigismond, Richard: 75
Sleeper, Raymond S.: 200
Smith, Wilbert: 7, 51, 53, 118, 175, 297-298, 299
Smith, Dr. (metallurgist): 138
Sperry, Maryann B.: 317
Stalin, Joseph: 304

Stanford, Ray: 188, 196-198, 199 (map), 204-205, 208, 210-212, 230, 317, 328, 340
Strickfaden, William: 72
Strugatsky, Arkady & Boris: 249
Sturgeon, Theodore: 249
Sturrock, Professor Peter: 52, 119, 120, 325
Sutherland, Kay: 14
Szilard, Dr. Leo: 310, 323

Teller, Dr. Edward: 60, 310
Truman, President Harry: 4, 14, 310, 323
Turner, Colonel: 33, 110, 328

Ussen (Apache God): 13

Vesco, Renato: 158
Vilchez, Ricardo: 121
Virk, Rizwan: 308

Vogel, Marcel (IBM Research): 299
Von Braun, Wernher: 160
Von Neumann, Dr. John: 60

Wade, Chuck & Nancy: 52, 315
Wagner, Lori: 77
Waldie, George: 75
Warren, Col. Stafford I.: 114
Webb, Walter (NICAP): 210, 211
Williamson, George Hunt: xvi, 322
Wilson, Admiral Henry Braid: 298
Wilson, Admiral Thomas Ray: 300, 301
Winker, Jim: 204
Wood, Ryan: 11, 136-138, 323, 341

Yost, Ed: 206

Zamora, Officer Lonnie: 41, 93-94, 207-210, 219-220, 230, 271, 328
Zinsstag, Mrs. Lou: 320

SUBJECT INDEX

140 Days to Hiroshima: 125

A Bodyguard of Lies: 181

Aerial Phenomena Research Organization (APRO): 119, 120, 325

Aermotor Company: 139, 327

Air force "Pogo": 201, 202, 203, 215, 303

Air Technical Information Center (ATIC): 180

Alamogordo, NM: xviii, 2, 14, (fig.8), 20, 33, 115, 190, 327,

Albuquerque, NM: x, xix, 2, 12, 48, 54, 59, 65, 76, 185, 188, 189, 195, 206, 289, 327

Aliens - see "creatures"

Alouette (helicopter): 216

Aluminum samples: 28, 45, 47 (foil), 48, 65, 86, 104, 109, 116, 117, 119-121, 125, 138-142, 162, 250, 278

Amazing Stories: xvii

Ames, Iowa: 310

American Institute of Aeronautics & Astronautics (AIAA): 132, 340, 341

Angel hair: 31, 35, 140, 165-167, 169, 240-242, 244, 246, 259, 273-278, 283-284, 297, 315

Anthropomorphic dummies: 159, 160

Apache People: 1&-13, 18, 63, 142, 318

Area 51 (Jacobsen): 12, 303, 304, 333, 337, 339

Arpanet: xvii

Artificial Intelligence: xvi, 123, 133, 178, 221, 271

Atomic bomb: xii, xiii, 4, 7, 19, 89, 113, 115, 151, 154, 158, 233, 240, 280, 286, 307, 310-311

Atomic Energy Department: 85, 89, 173, 187, 319

Atoms in the Family: 1, 309, 338

Aurora, TX: 116, 117

Austin, TX: xv, 121, 324, 340

Australian investigation: 78, 139, 317, 327, 332

Aviation Week: 179

B-25 aircraft: 20, 33, 327

B-52 aircraft: 181

Ball lightning: 205

Balloon hypothesis: xv, 19, 29, 32, 41-42, 64, 137, 142, 205-207, 216, 227, 259, 303, 340

Beliefs: 50, 52, 59, 82

Bhagavad Gita: 69, 233

Bible: xvi

BLM (Bureau of Land Mgmt): 11, 54, 78-79, 161, 173

Blue Book: see "Project Blue Book"

Bogota, Columbia: 120

Born on the Edge of Ground Zero: 12, 18, 27, 227, 309, 323, 331, 337

Bracket: xviii, 35, 42-43, 45, 65, 86, 92, 97 (Fig.20), 103, 105, 109, 125, 138-140, 141-142, 145, 151-152, 250, 256-261, 269, 306, 317, 327

Brain (effects on the-): 82, 128, 223, 224

Brookhaven, NY: xviii

Campamocha: 36, 37, 148 (fig.148), 146, 151

Campinas, Brazil: 118

Canada: 7, 50, 51, 297, 317-318, 326, 331

Canadian UFO Investigation Project: 7, 118, 297-298, 331

Cancer (caused by the bomb): 115, 258, 263, 282, 286-287, 309, 323
Catholic Church: 227, 236, 311
Cattle (effects on -): 17, 19, 80, 137, 287, 296, 332
CDC (Centers for Disease Control): 115
Centre National d'Etudes Spatiales (CNES): 215, 219, 226, 330
Chicago, IL: xii, xvii, 1, 177, 182, 185-186, 300, 321, 323, 325
CIA: 205-206, 231, 272, 294, 329-330, 335, 339
Clearances: see "security clearances"
Cockelbur: 64, 78-79, 80
Coincidences: 67, 75, 214
Cold War: 50, 86, 89
Computer technology: xvii, 87, 113, 138, 178, 183, 187, 234, 307
Condon study: 120, 197, 329
Congress: see "US Congress"
Contactees: see "UFO contactees"
Corona, New Mexico: 339
Corona virus: 103, 234
Council Bluffs (Iowa): 121-123
Crashes – see "UFO crashes"
Creatures seen:
- At San Antonito, NM: 21-23, 36-37, 43, 63, 98, 109, 127-128, 146
- At Aurora, TX: 116, 117
- At Socorro, NM: 188, 198
- At Valensole, France: 214-218, 222, 272, 307, 313, 316, 330
Czechoslovakia: xii, 159, 321

Dallas Time-Herald: 116
DARPA: xvii
Data bases (computer-): xv, 50, 81, 178, 196, 221
Datil, NM: 250

Day after Roswell, The (Corso): 300, 338
Dayton, OH: 127, 185, 186
Dearborn Observatory: 186, 202
Debris – see "UFO samples"
Delphos, KS: 63
DIA: 301
Diesel: 223
Diseases (from radioactivity): 115, 154, 237, 287
Dreams: 82, 99, 125, 128-129, 219, 225

Easter Bunny: 305
EG&G: 300, 333-334
El Bosque del Apache: 11
El Defensor Chieftain: 205
Emperor (of Japan): xiii, 5
Engine failure: 209
Enola Gay: 4
Evidence (from UFO crashes): xiv, 104-105

Fat Man: xiii, 4, 71 (Fig.15), 73, 152 (Fig.24), 156, 157
FBI: 118, 186, 188, 189, 190, 193, 195, 201
Feuerball: 159
Fiber "optics": 31, 138, 141, 168, 240, 241, 246, 250, 251, 259, 261, 269, 273, 274, 275, 282, 284, 288
Fire ants: 109, 111, 128, 147, 148
Flying saucer: xi, xvii, 33, 41, 132, 147, 149, 155, 159, 205, 225, 297, 298, 320, 321, 322, 338, 339
Flying Saucers: xvii, 331
Footprints: 190, 203, 222
Foreign Technology Division (FTD): 177, 179, 328
Fortean Times: 205

Subject Index

France: xii, 49, 119, 123, 214, 215, 219, 222, 330
French Army: 223
French Résistance: 215, 220

Gadget (The-): 4, 5 (Fig.4), 9, 73, 113, 114, 153
Gas diffusion: 7
Genealogy: 45, 85-87, 293, 337
Genesis: 159
Germany: xii, 16, 47, 158, 182, 187, 312
Gig Harbor, WA: 15, 45, 86, 327
Griffin Pipe Company: 121
Ground traces: 21, 27, 171-172, 180, 186-187, 195, 197, 199 (Fig.27), 202, 209, 214, 215, 219, 223, 226, 234, 313, 340
Ground Zero: xiii, 9, 11, 62, 67, 73, 85, 115, 130, 153, 154, 155, 157, 235

Habbebishopheim: 187
Hanford, WA: 7
Heavy water: 7
High-entropy alloys: 123
Hiroshima: xiii, 4, 14, 115, 151, 153, 155, 310-311
Holloman Air force Base: 233
Hollywood, CA: 131
Hombrecitos: 64, 112, 144
Hungary: xii
Hydrogen bombs: 154, 183, 323

I-Beam: 35, 140, 151, 162
IBM Advanced Research Labs: 299
Imaginal stage: 149
Information patterns: xx, 53, 81-82, 197, 213-214, 226, 340
Information retrieval: 178, 187

Insectoids: 128, 140, 147, 149, 225, 306
Insignias: 186, 190, 207 (Fig.28)
Institute for Advanced Studies at Austin: 121
Interference: 157, 158, 223
Internet: xvii, 44, 45, 85-88, 103, 212, 252, 277, 316
Iodine-131: 114
Iowa State University: 121
Italy: xii, xvii, 1, 49, 53, 88, 323

Japan: xiii, 4-5, 14, 76, 86, 115, 143, 154-155, 158, 182, 310, 312
Jerusalem cricket: 147, 148 (Fig.23)
Jopala, Mexico: 121-122
Jornada del Muerto: 2
Jumbo: 1, 6 (Fig.6), 8 (Fig.7), 9, 73, 152 (Fig.24), 157

Kiana (Alaska): 120
Kirtland Air force Base: 185, 190, 327
Kugelblitz: 159

Las Vegas, NM (mental hospital): 85, 94
Las Vegas, NV: 50, 85, 274, 300, 324, 331
Lawrence Livermore Labs: xviii, 333
Life Magazine: 312
Lightning: 13, 17, 46, 76, 91, 105, 106, 129, 144, 205, 238
Little Boy: xiii, 4, 155
Los Alamos, NM: x, xviii, 1, 2, 3, 7, 86, 92, 158, 241, 287, 296, 309, 311, 319, 327, 332, 333
Lunar landing module: 207

McChord Air Force Base: 228
McDonald Observatory: xv

McDonald Ranch: 9, 69, 72
McDonnell-Douglas: 116, 137, 326
Magnesium: 118-122, (Table 2), 139, 298, 325
Magnesium orthosilicate: 118
Magnetic waves: 158, 297
Majic Eyes Only: 11, 136, 323
Man-made UFOs 1944-1994: 158, 341
Manhattan Project: xi, 7, 60, 85, 95, 114-115, 157, 187, 297, 302, 323, 333, 337
Marconi Tower: 11, 20, 46, 62, 80, 90, 91, 140-141, 143-144, 160-161, 169, 171, 230
Mars (planet): xv, 50, 146, 231
Material testing: ix, 44, 53, 60, 212
MiG aircraft: 179
MK-ULTRA: 306
Maumee, OH: 119
Maury Island, MA: xiv, 117
Medical effects: 114, 115, 159, 232, 306-307
Memory effect (metal): 47, 240, 243, 250, 251
Memory (human): 162, 237, 260, 261, 265, 273
Metal samples: ix, xvi, xviii, 9, 15, 28, 29, 31, 35, 42, 45, 46, 47, 65, 79, 81, 92, 104, 105, 107, 116, 117, 118, 119-123, 128 (Fig. 22), 135, 137-143, 151, 156, 164, 180, 210, 212, 230, 235, 239-241, 243-245, 250, 253, 257, 258, 261, 264, 265, 270, 273, 298, 299, 315, 316, 327, 335, 340
Mexico: 12, 65, 121, 317
Mohave Desert: xvi
Mountain Mail: 142, 325
MUFON: 52, 77, 325

Nagasaki: xiii, 4, 14, 71, 73, 113, 115, 151, 153, 155, 310
NASA: xv, xviii, 180, 203, 210
NASA- Goddard: 210, 211, 212, 216, 330
Natl Inst. for Discovery Science (NIDS): 53, 300, 331-332
Nat'l Investig. Com. on Aerial Phen. (NICAP): 208, 210-212, 325, 329
Native Americans: 12-13, 42, 132, 232
Need to Know: 15, 44, 138, 211, 277, 324
Nevada Test Site: 115, 154
New Scientist: 131
New York Times: 133, 197
Nightmares: 83
Niño de la Tierra: 147, 148 (Fig.23)
Nitinol: 47
Normandy: xi, 181
Norway: 7
Northwestern University: xv, 179, 185-186, 299, 321
Nuclear chain reaction: 1, 7, 8, 323
Nuclear explosions: xi, xii, xvii, 2, 6, 8, 14, 19-20, 54, 69, 72, 114-115, 130, 153-154, 266, 287, 289
Nuclear Weapons Journal: 72

Oak Ridge, TN: xii, xviii, 7
Omaha (NE): 121
Oraison (plateau): 218, 219
Orsay University (France): 119
OUSDAT: 300
Owl Bar & Café: ix, xiii, xiv, xviii, xix, 31, 60, 78, 92, 93, 244

Palo Alto, CA: 131
Pan-American Highway: 62, 142
Paralysis: 218, 221

Subject Index

Paranormal Research Society: 53, 59, 318
Paris, France: xi, xii, 131, 215, 222
Pearl Harbor: 4
Peenemunde rocket base: 7, 312
Pentagon (The): 4, 115, 131, 153, 183, 241, 295, 328
Phoo bombs: 158
Photographs: 176, 180, 204, 209, 320
Plants: see "vegetation"
Plutonium: xiii, 4, 7, 8, 9, 14, 69, 71, 72 (Fig.15-16), 73, 113, 114, 154
Pontoise, France: xii, 107
Praying Mantis: 147-148 (Fig.23)
Pre-Astronauts (The): 160
Princeton University: 101, 320
Project Blue Book: xv, 173, 177-182, 185-187, 190, 193, 195, 197, 198, 200, 204, 209, 317, 322
Project Magnet: 7, 331
Project Manhattan: xi, 7, 60, 85, 95, 114, 115, 157, 187
Project Paperclip: 160, 312
Project Sign: 187, 188, 321
Project Grudge: 187, 188, 322
Puebla, Mexico: 121
Psychic impressions: 22, 37, 38, 39, 82, 99, 128, 129, 146, 147, 149, 151, 225, 307-308, 316, 334
Puget Sound: xiv, 117

Radar: 120, 132, 143, 159, 179, 299
Radioactivity: xiii, 7-9, 75, 85, 114, 115, 139, 153-155, 189, 209, 214, 240, 266, 276, 323
Retrieval schedule: 110 (Table 1), 243
Rice University: 149
Rio Grande River: 9
Roadside Picnic: 249, 302

Roswell, NM: x, xiii, xvii, xvi, 15, 16, 33, 37, 41, 47, 50, 88, 95, 136, 141, 143, 152, 159, 214, 233, 240, 303-304, 307, 324
Rowland Heights, CA: 15, 41

SAAB Corporation: 119
Samples: see "UFO samples"
San Angelo, TX: 187
San Antonio, NM: ix, xi, xiv, xviii, 3 (map), 9, 11, 12, 15, 24, 31, 44-45, 52, 53, 59, 77, 85-88, 94, 114, 130, 147, 155, 159, 161, 187-188, 214, 216, 225, 227-228, 236, 316
San Antonito, NM: 11, 13, 15-16, 62, 76 (Fig.17), 87, 93, 113, 116, 145, 154, 159, 223, 233, 235, 280, 293, 302, 304, 307, 313, 318
San Augustin (Plains of -): ix, 53, 68, 319
Sandia Mountains: 2
SAPOC (Special Access Program Oversight Comm): 301-302
Scanning Ion Mass Spectroscope (SIMS): 121
Scientists (lack of interest by -): xvi, 41, 42, 52
Seattle, WA: 15, 44, 138
Secrecy: ix, xv, 1, 26, 41, 89, 114, 153, 176, 232, 284, 294, 296, 309
Security clearances: xvi, 79, 89, 92, 126, 173, 177, 294-295
Silicon Valley: ix, 80, 131, 133, 141, 289, 295, 299, 309
Silumin (alloy): 97 (Fig.20), 101, 139, 140
Simulation Hypothesis: 234, 308
Sleep (disruption): 99, 190, 224, 226

Socorro, NM: xviii, 3 (map), 9, 41, 48, 62, 65, 78, 93, 110, 147, 183-191, 193, 195-212, 214-216, 226, 229, 230, 270, 271 (Fig.30), 272, 278, 303, 307, 317, 319, 325

Socorro Saucer in an Air Force Pantry: 188

Sounds: 14, 15, 19, 20, 22, 38, 91, 120, 146, 202, 215, 228, 230

Spacecraft: xvi, 226, 306

Spider web: 31, 167, 168, 172, 230, 235, 240-241

Stallion Site: 32, 57, 69, 95, 154, 156, 190, 208, 233

Stanford Research Institute: xvii, 352

Stanford University: xvii, 52, 119

Statistical studies: xv, 178, 181, 182, 214, 234, 332

Strategic Air Command (SAC): 181, 183

Synchronicity: 185

Tacoma, WA: 227

TNT: 8, 114

Traces: see "Ground traces"

Trailer (Army tractor-): 6, 9, 27, 30, 32, 35, 54, 106, 137-138, 141, 145, 156

Transistor: 47

Trinitite: 72

Trinity Site, NM: xii, 1, 2, 3 (map), 4, 5, 7-9, 11, 41, 60, 62, 69, 70 (Fig.14), 92, 95, 115, 146, 151, 159, 169, 228

Trinity, as a "test": 113, 151, 153

Trinity's Children: 113, 338

Truk Atoll: 187

Ubatuba, Brazil: 52, 119, 122-123, 325

UFO contactees: xiv, xvi, 320, 322

UFO crashes: ix, x, xiii, xiv, xvi, xvii, 16, 19, 33, 35, 37, 41, 49, 53, 61 (Fig.12 and 13), 82, 104-105

UFO "explanations": xv

UFO Investigator: 212

UFO samples: x, xvi, xviii, 4, 28, 60, 77, 118, 122 (table), 162, 178, 190, 209-210, 239, 241, 255, 258-266, 276, 299, 318, 325. See also "Aluminum samples", "metal samples"

University of Colorado: 120, 179, 197, 329

University of Texas: 121

Uranium: xiii, 4, 7, 323

US Academy of Sciences: 133, 197, 214, 294, 306

US Air force: xv, 147, 159, 173, 177-180, 190-202

US Air Force data-base: 187

US Army: ix, xii, 2, 11, 12, 13, 28, 29, 32, 33, 60, 64, 85, 103, 110 (Table 1), 136, 141, 157-158, 160, 243, 256, 267, 323

US Congress: xvi, 160, 181, 197, 213, 278, 294

US Atomic Energy Department: 85, 89, 173, 187

US Navy: 118, 298, 321

USS Missouri: 5

USSTAF: 158

USSR: 304

V-weapons (German): 158

Väddö island, Sweden : 119

Valensole, France: 214, 215, 216 (Fig. 29), 217-226, 271, 307

Vegetation: 54, 61-62, 68, 71, 76, 78, 80, 142, 186, 204, 225, 238, 318

Venus (planet): xv

Subject Index

Vertical take-off and Landing: 202

War of the Worlds (The): 304

Washington (State of-): xiv, 7, 15, 41, 48, 86, 117, 173, 227, 326-327

Washington, DC: 49, 118, 210, 278, 298, 299, 316

Weather (or hot-air) balloon: xv, 19, 29, 30, 32, 41, 42, 64, 137, 141, 143, 205-207, 216, 227, 259, 303

Westport, IN: 187

White Sands: x, xviii, 3 (map), 12, 28, 33, 68, 85, 113, 141, 145, 153, 159, 160, 190, 201, 206, 208-209, 230, 233, 282, 312

Windmill: 17, 28, 31, 44, 48, 116, 139, 141, 161, 263, 264, 327

World War Two: x, 50, 60, 86, 90, 158, 215, 225, 265, 274, 307, 316

Wright-Patterson Air force Base: xvi, 177, 186

Zimbabwe: 148

Zinc: 47, 65, 117, 139, 211, 212

Jacques F. Vallée

Born in France in a family with long international traditions, Jacques studied mathematics at the Sorbonne, earned a Master's degree in astrophysics at the University of Lille and was recruited to the first French team that tracked early artificial satellites at Paris Observatory.

Moving to the US in 1962, he pursued his passion for science working on NASA projects at the University of Texas in Austin (notably, coding the first computer-based map of Mars) before joining Northwestern University where he completed his PhD in artificial intelligence. Jacques continued his computing and entrepreneurial career at Stanford Research International and the Institute for the Future as one of the Principal Investigators on the early Internet before serving as a founder of a family of venture capital funds in Silicon Valley, specializing in information technology and biotech investments. The funds, including the first NASA venture fund, have financed 60 high-technology startups and led 20 companies to the public markets.

Dr. Vallée remains an active high-tech investor in Silicon Valley while serving on the scientific advisory board of the French Space Agency's group officially studying UFO reports. He has published his research in books that have been widely translated around the world. He was the real-life model for the character portrayed by François Truffaut in the film, *Close Encounters of the Third Kind*. Jacques lives between San Francisco and Paris. He has two children.

About the Authors

Also by Jacques Vallee
www.jacquesvallee.net

Novels:

FastWalker, novel (Berkeley: Frog, Ltd.)
Stratagem, novel (San Francisco: Documatica Research)

Science and Technology:

The Network Revolution (And/or, Penguin, Google)
Electronic Meetings (co-author, Addison-Wesley)
Computer Message Systems (McGraw-Hill)
The Four Elements of Financial Alchemy (Berkeley: TenSpeed)
The Heart of the Internet (Hampton Roads, free on *Google Books*)

UFOs and Paranormal:

Anatomy of a Phenomenon (Regnery, Ace, Ballantine)
Challenge to Science (Regnery, Ace, Ballantine)
Passport to Magonia (Regnery, Contemporary)
The Invisible College (New Yort: E. P. Dutton)
Messengers of Deception (And/or, Bantam)
The Edge of Reality (with Dr. J. A. Hynek, Contemporary)
Dimensions (Contemporary, Ballantine, Anomalist Books)
Confrontations (Ballantine, Anomalist Books)
Revelations (Ballantine, Anomalist Books)
A Cosmic Samizdat (New York: Ballantine)
Forbidden Science, Vol. 1-4 (Documatica, Anomalist Books)
Wonders in the Sky (with C. Aubeck, New York & London: Tarcher-Penguin)

In French:

Le Sub-Espace, novel (Paris: Hachette — Jules Verne Prize)
Le Satellite Sombre, novel (Paris: Denoël, collection Présence du Futur)
Alintel, novel (Paris: Le Mercure de France)
La Mémoire de Markov, novel (Paris: Le Mercure de France)
Les Enjeux du Millénaire, essay (Paris: Hachette Littératures)
Au Coeur d'Internet (Paris: Balland, free on Google Books)
Stratagème, novel (Paris: L'Archipel)
Science Interdite, Vol. 1 (Marseille: Parasciences)
Science Interdite, Vol. 2 (Genève: Aldan)

Paola Leopizzi Harris

Born in Rome in a family of Italian diplomats, Paola Leopizzi Harris studied English and French at Rhode Island College and earned a second Bachelor's degree in science at the University of Colorado. Her Master's degree in education from Leslie College in Boston led her to serve as the Principal for the American Overseas School of Rome. She taught ancient history, photo journalism and English there until she returned to the US in 2007 to pursue her interest in investigative reporting.

As a widely-published writer with a strong audience in Europe, Paola began studying unidentified aerial phenomena and was soon on personal terms with leading researchers of the field, notably as an associate of Dr. J. Allen Hynek in his investigations. She has published her interviews of numerous military witnesses in her widely translated books. In particular, she recorded the experiences of Colonel Philip Corso, whom she brought to Rome, and she wrote the preface to his book (*The Day after Roswell*) for the Italian public, as well as the preface to Apollo 14 astronaut Dr. Edgar Mitchell's book *The Way of the Explorer*.

Paola Harris has lectured extensively in the US and Europe about the importance of making ufology a serious Academic study, promoting international dialogue and open discussion of research data. Her current research focuses on events in Latin America and on related consciousness studies. She lives between Rome and Boulder, Colorado and has two children.

Also by Paola Leopizzi Harris
www.paolaharris.com

In English:

Connecting the Dots: Making Sense of the UFO Phenomena (Authorhouse)
Exopolitics: Gateway to a New Reality (AuthorHouse)
Exopolitics: All the Above (Authorhouse)
UFOs: How Does One Speak to a Ball of Light? (Kindle Publishing)
UFOs: All the Above …and Beyond (Kindle Publishing)
Conversations with Colonel Corso (Kindle Publishing)

In Italian:

Esopolitica: e Gia Nel Vento (Verdechiaro Edizioni)
Il Mistero Svelato (Verdechiaro Edizioni)
Esopolitica: Lo Stargate per Una Nuova Realta (Verdechiaro Edizioni)
Cosi Parlo Philip Corso (XPublishing SrL)

In Spanish:

Conversaciones Con El Coronel Corso (StarworksUSA LLC)
Esopolitica: Puera Estelar a una Nueva Realidad (Ediciones Obelisco)

In French:

Ovnis: Comment Parler à une Boule de Lumière (StarworksUSA, LLC)
Ovnis: Tout ce qui Précède… et Au-delà (StarworksUSA, LLC)
Conversations avec le Colonel Corso: un Mémoire Personnel et un Album

Made in the USA
Coppell, TX
22 December 2023